Digital
Desires

This book is dedicated to the memory of
Philippa Goodall
1943 – 1999

Philippa was a founder member of 'Cutting Edge'. She had been with the group since its initiation in the early 1980s and was a co-organiser of their first conference 'Women at the Cutting Edge' held at the Institute of Contemporary Arts in 1988.

She had been an active member of 'Cutting Edge' for over eleven years, travelling from Bristol where she worked as director of Photography at the Watershed, to attend meetings at the University of Westminster in London where 'Cutting Edge' was based. She was a co-organiser of the two subsequent conferences 'Desire by Design' and 'On/Off+Across', co-edited their first publication 'Desire by Design' and, most recently, worked as guest editor on their second publication 'Digital Desires'.

Philippa will be sadly missed both for her dedication to the work of 'Cutting Edge' and her immense contribution to feminism and to women's creative practice.

Digital
Desires

Language, Identity and New Technologies

edited by

Cutting Edge
The Women's Research Group

I.B.Tauris *Publishers*
LONDON • NEW YORK

Published in 2000 by I.B.Tauris & Co Ltd
Victoria House, Bloomsbury Square, London WC1B 4DZ
175 Fifth Avenue, New York NY 10010
Website: http://www.ibtauris.com

In the United States and Canada distributed by St. Martin's Press
175 Fifth Avenue, New York NY 10010

ISBN   1 86064 575 5

A full CIP record for this book is available from the British Library
A full CIP record for this book is available from the Library of Congress

Library of Congress catalog card: available

Typeset in Nimrod by The Midlands Book Typesetting Company
Printed and bound in Great Britain

CONTENTS

Acknowledgements

This book had its starting point at the conference 'On/Off+Across' which was organised by 'Cutting Edge' and held at the University of Westminster in November 1998. Subsequently 'Cutting Edge' commissioned a small number of additional papers which make up the collection now called 'Digital Desires'.

We are grateful to all those who originally participated in the conference and who have subsequently contributed to this book. 'Cutting Edge' is particularly grateful to Philippa Goodall for her work as guest editor and to Erica Matlow for her assistance in this task. We also want to thank Sandra Kemp for her work on the Introduction, Sarah Kember, Stevie Bezencenet, Jane Prophet and Jackie Hatfield for second reading and copy editing and Peter Kavanagh for his technical expertise in times of crisis.

'CUTTING EDGE'
THE WOMEN'S RESEARCH GROUP
AT THE UNIVERSITY OF WESTMINSTER

Foreword
About the Cutting Edge Research group

'Cutting Edge' is the women's research group at the University of Westminster. It includes women who work in and across art, design, media, architecture, photography, language and cultural theory. The group, at present, consists of ten women who work either full or part time at the University of Westminster, Stevie Bezencenet, Helen Coxall, Sandra Kemp, Jackie Hatfield, Erica Matlow, Marion Roberts, Gail Pearce, Jane Prophet, Helen Reddington, Alexa Wright and with associate members from City University, Goldsmiths College, de Montfort University, and Napier University.

Cutting Edge is committed to a feminism of empowerment which acknowledges the many complex issues that still need to be fought for in relation to the rights of women. Being on the 'cutting edge', we believe, enables us both to empower ourselves and a wider audience. We also believe that through activities like conferences, books and exhibitions we can create a space where it is possible to participate in, and to critically discuss, the relatively early stages of technology in order that the field is not completely dominated by patriarchal ideologies.

Introduction

Feminist Women have a long history of dancing through a variety of potentially lethal mine-fields in their pursuit of socio-symbolic justice. Nowadays, women have to undertake the dance through cyberspace, if only to make sure that the joysticks of the cyberspace cowboys will not re-produce univocal phallicity under the mask of multiplicity, and also to make sure that the riot girls, in their anger and their visionary passion, will not recreate law and order un-der the cover of a triumphant feminine.

Rosi Braidotti, 'Cyberfeminism with a Difference', 1996

The past ten to twenty years have seen sweeping changes in the established and newer physical and social sciences, often accompa-nied by closer association between hitherto discrete disciplines. A panoply of technologies and scientific procedures are the vehicles for such developments, enabling new channels of enquiry to be pursued and long-standing hypotheses to be explored, challenged and either brought into the scientific fold or excluded from it.

Digital Desires examines the ways in which late twentieth century 'increasingly capacious' notions of the self are being reconfigured and how new technologies play a part in producing new positions for identity in the twenty first.[1] Different contributors using a variety of media assess the impact of technology on our un-derstanding of 'production' and 'reproduction', 'space' and 'communication' and 'language' and 'identity'. Genetics and bio-engi-neering have changed conceptions of the basic components of carbon based life and taken the awe-inspiring step of crossing the species boundaries for both experimental and economic advancement. Such 'advancements' are defined in relation to plant and animal life, and increasingly in relation to human life. At the same time the biologi-cal and computer sciences are being harnessed to produce virtual 'life' forms that exist only in an electronic environment. The production and reception of discourses of body and identity are also central to the feminist project.

INNERTEXTUALITIES
CYBERBODIES AND PSYCHOANALYSIS
The first section of the book focuses on an exploration of the impact of a number of technological developments – imaging and medical technologies, bio-technology and artificial life – on what have until now seemed to be relatively stable (if contested) concepts of the body and the self.

Each chapter suggests either implicitly or explicitly there is a pressing need to consider the volatile implications of the breaching of boundaries between the physical and the virtual body, and the different registers of lived versus virtual experience. Psychology and psychoanalytic discourse are mobilised to interrogate and illuminate the products and processes that are being spawned in science, culture and industry and to argue for caution in the current somewhat utopian climate of reconceiving the praxis of physical, intellectual, emotional and social life.

Thus, Sandra Kemp asks whether technology may be the key to a new kind of physiognomy and considers the effects of new technology, such as digital imaging or Maxillofacial Surgery, on the psychology of face recognition and portraiture. Jane Prophet and Sian Hamlett use psychoanalytic theory to analyse the CDROM artwork "The Internal Organs of a Cyborg". The female protagonist in this artwork uses implant technology and surgical augmentation to test her physical boundaries. In asking the question 'whose reality is it ... ?', Tessa Adams explores a confrontation between the notion of an 'authentic' reality and the territory which is deemed 'virtual'; both, she argues, are increasingly problematised through technological discourse and practice and in the growing dislocation between lived and enacted 'identity' as it is played out in the virtual environment of the internet.

In the final chapter of this chapter Sarah Kember asks how ALife deploys gender and identity, and what the values of artificial societies will be and who will define them.

OUTERTEXTUALITIES
INTERNAL SITES AND EXTERNAL EXPERIENCES

Throughout *Digital Desires* optimism about the potentialities of the new technologies is tempered by urgent questions about the relationship between the new technologies, the arts and militaristic and personal surveillance. New media technologies are changing relationships between audiences, artefacts and galleries resulting in what Rosi Braidotti describes as 'the new and potentially fruitful alliance between technology and culture'.[2]

Technologies of production have not only radically changed working conditions and possibilities for both art and artist but have also affected our subjectivities and fantasies. The essays and images in the second section of the book set out to unravel a complex web of relations between conceptual and discursive formations about virtual and 'real' words and worlds. Not all technologies discussed here are new, however. Indeed older technologies often constitute the 'outertextuality' referred to in the title of this section – an invented word that plays on its obverse actual word 'intertextuality'. The older technologies have become part of the 'naturalised' environment that informs and partly determines new technologies and the ways society engages with them.

In the first paper in this section, Sherry Millner examines what have become the conventional cultural technologies of the late twentieth century, video games, television and aspects of military technology that have been in existence since at least the Vietnam War. These, coupled with the domestic site of the American home, form the context for her search to understand the driving forces behind the appalling phenomenon of juvenile murderers in 1990s' America. Angela Medhurst examines the social implications of electronic supermarket. E-business and e-shopping are becoming more evident in the online environment and e-food shopping is under trial by a number of supermarket chains.

Still in the virtual world, Maren Hartmann traverses the virtual boulevards of cyberspace and considers the gendering of new online social positions for identity. Taking the cyberflâneuse as her case study, Hartmann argues that femininity remains as problematic a category in the virtual world of the urbanised late twentieth century as it was in the physical cityscape of the late nineteenth century.

Meanwhile joint research by Penny Harvey and Gaby Porter looks at the explicit and implicit gender relations of attitudes and working practice, in the material environment of both physical and virtual 'space'. In 'Windows on the World', the penultimate chapter in this section, the social relations of both material and conceptual space are examined. Jos Boys argues that much contemporary cultural and architectural theory has a tendency to close down, rather than open up, spaces for thinking about material space, architecture and identity. Stevie Bezencenet's visual essay 'Frontier Dreams I', which ends this section, avoids closure by extending the examination of space into 'the technological sublime'. Her female protagonist inhabits a brave new world of technological advance.

INTERTEXTUALITIES
LANGUAGE, IDENTITY AND NEW TECHNOLOGIES
Issues of translation, transformation and intertextuality surface throughout *Digital Desires* as work gets re-presented and thus recontextualised as it is placed in different media. The essays in the third and final section of the book share a preoccupation with language, dealing with personal artistic and work-life practice in the form of theorised narratives. The authors invite interactivity between a combination of web-authored and written work.

Whereas in Section One Prophet and Hamlett played with images and concepts of new technology that some might argue represent the self-absorbed urban west, in the third section Lucia Grossberger-Morales uses the computer in her artwork in order to focus on her own relation between the so-called third and first worlds. Bolivian by birth and American by adoption, for this artist the computer has offered the means and space to speak of otherness.

As a musician Helen Reddington explores the options open to her and other women in pop music. Her chapter explains the thinking behind and making of Voxpop Puella which consists of six films and a live performance which explores attitudes to the seven ages of women, using predominantly digital film and music technology.

Erica Matlow, meanwhile, explores some aspects of the relationships people have with their computers by carrying out a small informal investigation through which she considers the impact for twelve women of their introduction to and use of computers. Whereas the response of these respondents in 'Women Computers and a Sense of Self' is largely positive about their sense of identity in their encounters with computers, Jackie Hatfield focuses on computer coding as the most significant element in the production of meaning in/for computing. She counters the arguments around technology by reminding the reader that the internet and computers do not in and of themselves change the ideologies of gender; and in her chapter wonders whether 'the hegemonic patterns that exist in real space will be paralleled in virtual space'.

The articles in *Digital Desires* demonstrate the importance of technology and its use as a cultural framework. They examine the ways in which new technologies are having a radical impact on existing debates about gender – and throw up new ones. Above all, however, they demonstrate feminism's political commitment to diversity.

Sandra Kemp

References
1. Helena Michie, 'Not One of the Family: The Repression of the Other Woman in Feminist Theory', in Marlene Barr and Richard Feldstein (eds.), *Discontented Discourses: Feminism/Textual Intervention/Psychoanalysis.* Champaign: University of Illinois Press, 1989, p.43.
2. Rosi Braidotti, op.cit., p.9.

Section 1
Innertexualities
Cyberbodies and Psychoanalysis

Technologies of the Face

Sandra Kemp

Writing on *The Human Face* in 1974 John Liggett announced that:

> *The time is not too far distant when the questions linking face and character will be written with certainty – when we shall be able to relate truly accurate descriptions of facial forms, textures and movements with genuinely objective movements and analyses of personality. We must await the development of an entirely new technology of facial description and analysis which matches in power and precision the science of personality. Then we will be able to pronounce with certainty on the relationship of face and character. Then we shall know whether and to what degree certain aspects of facial structure and movement are serviceable guides to personality and experience.*[1]

In the twenty-five years since Liggett was writing, there have been significant technological advances in the description and analysis of facial structure and movement. Facial perception and recognition are now a credited field of study within psychology, medicine (particularly forensic science and cosmetic surgery), philosophy, semiotics, cultural theory and experimental art practice (for example, digital art). Drawing my examples from exhibitions, from the Facial Image Archive, from recent scientific research (by Richard Kemp, Nicola Towell, Graham Pike, Alf Linney, Vicky Bruce and Andy Young) and from literature, film and photography, in this chapter I will look in particular at the overlapping discourses of new technology, science and art. In so doing, I will consider how the new facial topographies made possible by new technologies affect our sense of 'self'. Is technology the key to a new kind of physiognomy that will unravel the relationship (or lack of it) between personality and character in the presentation of the face? What is the relationship between the conceptual and the phenomenological, between the self, the face and art?

The first image was a face. In classical mythology a lovely youth named Narcissus lay beside a pool gazing in adoration at his own reflection. Ignoring the loving attention of the nymph Echo, he wasted away, died and was metamorphosed into a flower bearing his name.[2] History is populated by faces: from Narcissus to Marilyn Monroe, immortalised in Andy Warhol's pop portraits of the 1960s, such as his huge 'Marilyn Diptich', where fifty identical images of the actress-celebrity, slightly modified in colour and tone, are deliberately shown together in a single work; from Saint Veronica, who, according to the Bible, compassionately pressed a cloth to Christ's face as he stumbled to Calvary, and found his image miraculously imprinted onto the material, to Anthony Noel-Kelly's notorious gilt-sprayed cast head of an elderly man caught in the grimace of rigor mortis, and from family photo-albums to police mug-shots, where, as Alan Sekula points out, there is a linkage between the spheres of culture and of social regulation: 'every proper portrait has its lurking inverse identification in the files of the police'.[3]

In 1997 two films on general release centred spectacularly on the relation between identity and faces. In the first, *Face-Off*, directed by Nicholas Woo and starring John Travolta and Nicholas Cage, FBI agent Sean Archer undergoes radical laser face surgery allowing him to switch faces with the comatose terrorist Castor Troy and assume his identity. In the second film, *The English Patient*, (based on the novel by Michael Ondaatje), after a badly burned pilot (Ralph Fiennes) is pulled out of the wreckage of his plane in the Sahara desert, he is placed in the care of an army nurse (Juliette Binoche) and identified only as 'the English patient': 'A man with no face. An ebony pool. All identification consumed in a fire ... There was nothing to recognise in him'.[4]

'Every age has its own gait, glance and gesture', wrote Charles Baudelaire in *The Painter of Modern Life*.[5] Writers and artists (as well as cultural and art theorists) have argued that the history of modern self-identity and subjectivity is inseparable from the human face.[6] Portraiture established itself in the Renaissance, but the genre existed even in antiquity and the early Christian world in the shape of sarcophagi, statues, busts and coins. Portraiture is a form of representation that is especially attuned to change, and in particular to changes that encode the ideas that generate identity. In *Immortality* the novelist Milan Kundera noted: 'You know me by my face, you know me as a face, and you will never know me in any other way. Therefore it could not occur to you that my face is not my self'.[7] For the photographer Philippe Halsman fascination with the human face determined his work: 'Every face I see seems to hide – and sometimes, fleetingly, to reveal – the mystery of another human being ... Capturing this revelation became the goal and passion of my life'.[8] The artist Stephen Farthing concurs: 'People talk about capturing the essence of a person. What I want to record is the topography of a face. I want to crawl all over a person's face and find out how it works'.[9] The academic John Hull, who went blind at the age of 43, describes how in the year or so since going blind all sorts of visual memories had flooded his consciousness, but the memories of faces had been the most important of all: 'To what extent is the loss of the image of the face connected with the loss of the image of the self?' he asks. He goes on to talk of '... the horror of being faceless. Of forgetting one's own appearance, of having no face. The face is the mirror image of the self'.[10]

Today representations of images of faces are part of the currency of everyday life. We find them on stamps, coins and banknotes, and on passports, identity cards, bus passes and library tickets. Pictures of writers and celebrities are used in advertisements and on book covers. Caricatures and cartoons are as popular as ever. Portraits also appear among boardroom paintings, drawings by street artists, and, of course, family snapshots. However, as the neurosurgeon Jonathan Cole points out, in a collection of interviews published earlier this year, the medical profession has little awareness of facial expression and its importance. Cole writes about the relation between facial mobility, intelligence and understanding; how facial problems might affect social development and selfhood. In *About Face* he interviews people whose faces have altered their lives. His first case study, Mary, is of a woman who suffered facial paralysis through a stroke, and then wasted away because she could no longer communicate:

'Without a face she as a person was all but invalidated,' Cole remarks. The other subjects of his study also reveal how the immobility of the face affects the experience of emotion itself and on how much our well-being depends on an emotional sensibility revealed to others via the face:

> *'The case histories and narratives in this book ... tell of difficulties in the calibration and experience of emotion and in the deeply embedded role of the face in our perception of self. ... All tell of the essential role of the face in the expression and experience of feeling itself. These who were hardly aware of the facial origin of their problems show how deep within us are these matters that they are only brought to light by a shattering disconnection between personality and the face.'*[11]

Here I want to raise questions about these issues around two examples taken from recent exhibitions linking technology and faces: the first in London in June 1998, and the second in Tokyo in November 1998.

'10'

Matthew Cornford and David Cross's '10' was one of six short-listed pieces in a competition for new digital art shown at the Institute of Contemporary Art in London.[12] Cornford and Cross had put up posters in Derby advertising a 'unique beauty contest ... looking for the face of Derby pride'. Almost 200 people turned up to be photographed (straight on, harsh light), and then have their image judged by a computer. 10 winners were selected and put in order according to the 'percentage of their beauty'. The finished artwork consists of the winning mugshots blown-up billboard-big. 'When we had the final selection of faces, they seemed bland,' said Cross. 'Some of the women, in particular, looked so similar as to be easily confused.'[13]

Why is this? Because scientific research suggests that symmetry – the classical requirement of beauty advocated by everyone from Plato onwards – remains the one aesthetic constant, Cornford and

Cross programmed the computer to pick out the most symmetrical faces. They used technology developed for digital 'fingerprinting' and also gave the computer a control: a kind of androgynous uber-beauty created by melding a range of famously beautiful faces, from Nefertiti to Keanu Reeves.

The computer then took 24 separate facial measurements from each contestant. There is almost something sinister about the chosen images, their uniformity reminiscent of those Nazi photographs of pristine Aryan specimens in which their facsimile features make them appear like androids. In contrast, when the Nazis photographed the 'rejects' in the concentration camps, the photographs were full of life in all its individual and quirky diversity (i.e. not symmetrical).

Coincidentally, at the same time as the ICA exhibition, Piotr Uklanski's controversial exhibition 'The Nazis': 122 pictures of actors in Nazi uniform, lifted from films and reduced to identically sized head-and-shoulder portraits, was showing at the Photographer's Gallery in London.[14] Critics were divided between those who thought Uklanski glamorised Hitlerism and those who saw the exhibition as a sombre bitter joke about the classificatory power of the photographic image and the cultural logic of fascism. Should artists reproduce a fascist aesthetic? Is this more than a mockery of Hollywood's representation of the War? Uklanski never answers the question implicit in his exhibition. But Uklanski's faces, like those in Cornford and Cross's '10', do make his audiences ask what technology does to faces, and what we do to/with technology? Cornford and Cross regard their exhibit as an intervention: 'People are afraid of new technology and what it will do to our society. *We want to question what our society will do to technology*: it is our own values – our anachronistic belief in physical perfection that will shape the way the future develops.' (My italics.)[15]

Similar issues are picked up by Mongrel, a London based arts group, whose 'Natural Selection' is an attempt to curb the activities of online racists. 'Natural Selection' works like any other search engine, such as Yahoo or Altavista, but is aggressively anti-racist. If you attempt to visit racist sites, you are confronted with the rotating eyes of a black person's head. Attempts to quit are greeted with a series

7

of threatening pop-up boxes: 'You can't get rid of me.' The more you try to escape, the quicker the eyes rotate, and the dialogue boxes reappear. Eventually it allows you to quit, but ten minutes after you've disconnected from the Internet, a black woman's face reappears, rotating on your screen again and refusing to go away. It's as if your computer has been infected by a virus. Just an arty prank? Matthew Fuller denies this:

> This is not a practical joke. What we are trying to do is to get people to think about the information they have access to on the net and *challenge accepted ideas about the role technology and classification theories play in upholding the status quo* round the world. (My italics.)[16]

'Classification theories' and science/technology have always played a significant role in characterising the face. From the ancient Greeks to Cornford and Cross, the classification of the beautiful (and 'good' face) is based on balance and symmetry. Recent research has apparently demonstrated that we are genetically programmed to prefer symmetrical faces. According to the evolutionary psychologist Steven Gangestad, 'Symmetry alone explains why Elizabeth Taylor, Denzel Washington and Queen Nefertiti are universally recognised as beautiful'.[17] Hence, although there are wide differences between individuals and across cultures in what is considered attractive, there is also a remarkable consensus of agreement about what constitutes beauty. For the ancient Greeks the idea of facial beauty was a question of proportion; mediaeval artists believed the perfect face was neatly divisible into sevenths. Even as late as the eighteenth century Sir Joshua Reynolds still believed that beauty was simply a matter of physical proportions. Even Hogarth believed he had found the overriding principle in 'the wavy line of beauty', 'the greatest, indeed the indispensable element of all beautiful things is the smooth serpentine line'. In 1990, in All Saints' Church in Newcastle, the performance artist Orlan began a project which involved using computer software to create a new self-portrait in which her own features blended with the chin of Botticelli's Venus, the forehead of Mona Lisa, the eyes of Gerome's Psyche, the mouth of Boucher's Europa, and the nose of a

Diana from the school of Fontainebleau. Orlan's 'performance' involves undergoing surgery to reshape her face into this new image: to-date she has had nine such operations, some of which were broadcast live to various locations. When Orlan's face was prepared for one of these operations, the markings were reminiscent of medieval and physiognomic texts.

Obviously these shared characteristics arise through striking evolutionary resemblances across cultures in the ways that different human faces are structured and grow. Even today our understanding of faces is influenced by the subtle interaction between social and artistic conventions and associations that have developed through history and which have generated stereotypes and prejudice associated with the face, and, in particular, to stereotypes suggesting some kind of moral congruence between inner and outer. 'I'm determined to read no more books where blond-haired women carry away all the happiness', announces Maggie Tulliver in George Eliot's *The Mill on the Floss*.[18] Writing in his *Guardian* 'Diary', the columnist Simon Hoggart notes:

> *I used to know a spectacularly ugly man who was nevertheless a great hit with women. To those people who found a form of words to ask him, without being too offensive, how he managed it in spite of his physical disadvantages, he would say: 'Give me twenty minutes and I can explain away my face.'*[19]

Research has shown that we are highly susceptible to quite small changes in the detail of a face, and how different emotional states result in changes to facial musculature. So, a slight increase in the distance between the eyes of a person causes that person to be judged more honest and reliable. If the distance of the brows from the eyes is judged to be too close, they create a falsely 'malevolent' expression, and too high, they suggest open-mindedness and accessibility. Plastic surgery which changes facial features creates a different character. Although advanced technologies of facial repair can be taken to extremes. In the *Village Voice* in October 1991, Michael Jackson's face was described as follows:

> *It's been planed and sheared, deconstructed and deracinated. Molded for the screen, taut for the billboard – Michael Jackson's face has become its own simulacrum. There is no original. He must startle himself when he looks in the mirror. He is pure image ... After an estimated nine rounds of plastic surgery, and a chemical face peel to lighten his skin ... he wants to be raceless and genderless and ageless – Mr. Postmod. ... the embodiment of artifice upon artifice.*[20]

For the most part, however, in the 80s and 90s, technologies (of medicine and of art) have increasingly worked to contradict the assumptions of universality that concepts of beauty have historically implied. The study of face perception has been greatly enhanced in recent years by the transformations that can be made using computer graphic techniques (especially morphing, or, computer produced manipulations of images in which the difference between two or more faces is reduced to achieve a blend of one (or more) faces into another). The new scope is seen in daring new facial iconographies of even conservative subjects like the Queen in which the ideological role of art in normalising and policing as well as creating facial images is interrogated. Consider, for example, 'The Morphing of Queen Elizabeth' (1994) which uses the computer program morph to produce images of the head of state as a woman of colour, or Justin Mortimer's (1995) royal portrait, in psychedelic colours, with the queen's head floating above her body. Or the website 'Fa(e)ces of the World' which promises 'hours of entertainment mutilating the images of the rich and famous'.[21]

One of the advantages of new technology has been to highlight the important role of movement in face recognition and in the perceptual processes underlying face recognition. In this context it is interesting to surmise what may have been lost from the absence of movement in early photographs and portraits. At the University of Westminster, Richard Kemp, Nicola Towell and Graham Pike's Face Processing Research Group has worked on the effectiveness of photo-credit cards, the use of dynamic mug-shots, witness identity masking, a comparison of face perception skills of young and old children, the electrophysiological evidence of the specificity of face perception, the separate impacts on face perception of hue and brightness changes, and the use of E-FITS.[22] Meanwhile, related work on face perception by Vicki Bruce and Andy Young at the Universities of Stirling and York has also demonstrated that face recognition is part of a whole perceptual framework of sensation and complex mental processes. Their most recent book, *In the Eye of the Beholder: The Science of Face Perception,* accompanied an exhibition on 'The Science of the Face' at the Scottish National Portrait Gallery in Edinburgh in 1998. Bruce and Young explore the huge range of experimental data on face processing (on how we recognise faces); why, for example, upside down faces are so hard to recognise and what makes some faces more memorable than others. Face recognition seems to depend on quasi-artistic effects: the correct balance of relative brightness, and context. Apparently our preference for seeing faces is so strong that it can over-ride the usual assumption that light comes from above. Over the past decade the growing presence of closed circuit television in shops, banks and public spaces has become accepted in the courts as a powerful identification tool to back up, or even replace, eyewitness accounts of incidents. According to Vicky Bruce such confidence is almost certainly misplaced. Changes in viewpoint, expression, lighting (and of course hairstyle and glasses) makes recollections of a face notoriously unreliable. Because we have not had time to build up a composite picture of a stranger, we have to rely on the brain's ability to initially perceive objects only in terms of light and shade, particularly shade. Bruce explains:

> *So even when the lighting is changed we find it diffi-*
> *cult to verify whether it is the same person, even when*
> *given limitless time ... Lighting is a very complicated*
> *thing that has been neglected. It is really easy for it to*
> *change how we see images and facial features. Both hu-*
> *man vision and computer vision find light and shade*
> *hard to deal with.*[23]

Images captured on security cameras are often of extremely poor quality, and thus much interpretation is involved in determining to

11

whom a captured face belongs. Advances in the methods that can be used to prove identity in these cases must await further research into the best way to measure and compare faces.[24]

Bruce and Young's findings are also examined within a wider evolutionary context in which the utility of being able accurately to recognise the individuality of a face and read moment-by-moment changes in it are cogently explored. A photo of Mars taken in 1976 by the Viking spacecraft would seem to display the natural land formation in the shape of a clearly imagined face. A photo of the same area taken twenty-five years later by the Mars Global Surveyor spacecraft shows nothing – no face. How do we account for the difference? And why do we see a face in the first picture? According to the scientists it is a conjunction of a trick of the light and some fantastic neural machinery that evolution has bestowed on us. We see a face in the Viking Cydonia image because there are more areas (more correctly, volumes) of nerves in the right hemisphere of our brains which are specialised to pick up the distinct shape of human faces: oval shape lit from above, eyes at top of nose, mouth below nose, chin. Why have these developed? Because it is a survival characteristic for babies to be able to recognise human faces over other species, since they rely absolutely on humans for their survival. From a very early age we can recognise faces in any orientation. We can infer the presence of a face from just a few visual clues. Later on, it's helpful, but not essential, for adults to be able to pick out human faces from vague shapes. At University College London Alf Linney is researching into whether faces could be used for identification in the same way that fingerprints already are. He takes 3D scans of people, and marks the images at points such as the corners of the eyes and the mouth. He can then describe the faces in terms of distances and angles between these marks. He hopes that one set of these values will turn out to be unique for every person.[25]

Critics of Linney's theories argue that too much of our appearance is dictated by influences that are not reducible to genes. It is also the case that the stereotypes and prejudices generated by the face have arisen from what are now regarded as debatable nineteenth century scientific theories (not least because of their implicit eugenics). The first of these is, of course, physiognomy, said to have begun with Pythagoras, whose views are developed in *Physiognomica* (sometimes attributed to Aristotle). The notion that a person's true character is revealed in his or her physical appearance has remained the striking concept for the western imagination ever since. And not only classical antiquity, but all the ancient cultures (Egypt, Arabia and China) value the study of physiognomy. From the eleventh to the eighteenth century physiognomy was bound up with both medicine and physiology. But in the nineteenth century physiognomy became dominant again and theories of physiognomy dominated culture and literature.[26] Its main proponent, Johann Kaspar Lavater (1741-1801) is largely forgotten today, but in his time he enjoyed the kind of adulation bestowed on film stars and pop idols.[27] The other nineteenth century 'face' sciences were Phrenology (the 'science' of looking at the skull as a source of character revelation), Darwin's evolutionary theories and Lombroso's criminal anthropology. All three work from the extrapolation of human difference to theories of rigid limits, what Gould calls: 'The tenacity of unconscious bias and the surprising malleability of quantitative data in the interests of a preconceived idea.'[28]

In *Criminal Man,* 1876, Lombroso derives his evolutionary theory of hereditary criminality from looking at a skull: 'At the sight of that skull [the brigand Vihella], I seemed to see all of a sudden, lighted up as a vast plain under a flaming sky, the problem of the nature of the criminal – an atavistic being who reproduces in his person the ferocious instincts of primitive humanity and the inferior animals.'[29] Examining and collecting skulls was a Victorian hobby and 'Phrenology' was still fashionable in the early twentieth century. In its January–June issue 1901 *Strand Magazine* published an article by Gertrude Bacon which (in all seriousness) attempted 'to apply our knowledge to the solving of that all-important question, "Has Baby a clever head?"'[30] Lombroso also postulated a meaningful similarity between the facial asymmetry of criminals and flatfishes with both eyes on the upper surfaces of their bodies. Darwin too sought to understand the semiotics of facial expression by comparison with animals (on one occasion he visited a zoo and put a freshwater turtle in with the monkeys to see if they expressed surprise). He wanted to discover why certain movements communicate certain expressions.

Half a century later the art historian E.H. Gombrich was still considering human facial expression by comparison with animals:

> *Unless introspection deceives me, I believe that when I visit*
> *a zoo my muscular response changes as I move from the*
> *hippopotamus house to the cage of the weasels ... this doc-*
> *trine relies on the traces of muscular response in our*
> *reaction to forms; it is not only the perception of music*
> *which makes us dance inwardly, but the perception of*
> *forms.*[31]

All these theories are heavily dependent on Lavater's physiognomy: the belief that character was indicated by facial features. In the mid-nineteenth century, physiognomy was offered as the new science of diagnosis, like the stethoscope, or chemical analysis in medicine, as part of a true 'science of character'. That art should look to science, both for specific information and for a methodology which would assist the artist in his conscious attempt to attain a convincing image of nature was accepted as a natural outcome of living in a scientific age. The history of the interplay of art and science in theories of the face is crucial. Throughout the late 19th century the link between criminal anthropology, art and physiognomy was vital. As a result, the Victorians were expert face readers and were preoccupied with faces. There were very different habits of observing and describing the face in the last century: every feature had a specific meaning: the nose was supposedly the indicator of taste, sensibility and feeling; the lips of mildness and anger, love and hatred; the chin, the degree and species of sensuality; the neck, the flexibility and sincerity of the personality, and so on. As Mary Cowling points out, we no longer scrutinise, record every detail and deduce the character in the same way. 'In the Victorian era art and illustration played an essential part in the public exchange of ideas and assumptions about human types providing a means of visual embodiment and helping to give credence to them with each example.[32] A large number of such books were aimed at the popular market and had a strongly practical orientation, for example, W.H.Hatfield, *Face Reading: with Hints on Love, Courtship and Marriage* (1870).[33] Francis Wey, visiting England from

14

France in the 1850s, records the advertisement he saw for a pantomime, 'The Prince of Pearls', at the Surrey Theatre which boasted that, as advisers to the production, specialists of every category had been consulted, archaeologists, zoologists, phrenologists and physiologists: the phrenologists having assisted in the making of masks scientifically moulded of the famous characters appearing in the show.[34]

Physiognomy was superseded by the 'hard' sciences in the early twentieth century, but the recognition and classification of facial images is still a subject for debate and one in which technology is playing an increasing role. In 'The Body and the Archive' Alan Sekula examines the history of police files – the cataloguing and classification of faces for the purposes of criminal identification.[35] The police 'mugshot' is now a familiar image of everyday life. And such images are currently used equally for the purposes of identification as well as arrest (on library cards, driving licences, passports and so on). However, the issue of facial ownership remains tantalisingly unclear. If the representation or image of a face may be used, it cannot be contained or constrained by the law, as is demonstrated by the unsuccessful attempts by the estates of Elvis Presley and Diana, Princess of Wales to patent their faces in 1998. (Interestingly, having failed to get 26 'representative' faces of Diana patented, her estate is now attempting to patent a virtual Diana made out of a composite of the twenty-six images.)[36] Marcus Harvey's 'Myra' is a fascinating example of these issues. 'Myra' is a reworking of one of the most memorable photos of our time, the tabloid's favourite Myra Hindley photo, made by using a template of a child-sized hand-print hundreds of times to build up an enormous digital image (9ft by 11ft). Hindley herself protested from prison about the use of her face on this image (and its alleged effects on her applications for parole). In 1997 the vice squad were called to the Royal Academy's 'Sensation' exhibition to consider 'a reasonable basis for prosecution' with respect to the portrait; none was found. And then, in a marvellous ironic twist, when Myra was subsequently attacked and defaced, the police were then called in to 'protect' the portrait.[37] Two years previously on its July 27, 1994 cover, a notorious issue of *Time* magazine featured a computer 'photo-illustration' of O.J. Simpson's mug-shot by Matt Mahurin that darkened his skin, blurred his features and thickened the stubble on his chin. The subsequent image was seen to reflect 'the predominant view' that Simpson had appeared 'to many whites [like] a fellow white man whose face grew darker as the proceedings went on'.[38]

So, who owns the representation of a face, and what is the relationship of that to the real person? Questions of ownership also arise with respect to facial images of well-known figures who become (as it were) national treasures. Two recent examples of this are the celebrity make-over of Jesus from 'an effeminate fairy in a white dress' to Che Guevara, and the ongoing debate over the Rice portrait of Jane Austen (considered by many to-date as the most authentic representation we have of her). Reviewing the current debate about the authenticity of the image, Claudia Johnson asks:

> *How could any image be commensurate with what we think and feel about Jane Austen? How could so momentous and yet so intimate a figure, one who has become the site of such different and such contradictory visions and fantasies – about civility, love, history and England itself – survive figuration without arousing disappointment and anger?* [39]

Electronically Yours

My second example sets out to question whether the combination of technology and science might provide an avant-garde portraiture which could tell us something new about the subject. Entitled 'Electronically Yours', Jasia Reichardt's exhibition at the Museum of Photography in Tokyo gives us quite new images of the human face and body.[40] Techniques include holography, video, interactive computer systems and digital animation. Faces are at the core of the exhibition. Some of these animated portraits speak and introduce themselves; some gradually change into a cake, with a streak of orange icing for the smile. 'Dissolution' is based on continuous transformations of the face. Paintings and drawings of heads, mainly of Christ, by Bellini, Massaccio, Botticelli and others are seamlessly transformed from one to another with quite new (and occasionally sinister) expressions appearing. Other exhibits transform the face from young to old, and give us a feel of what constitutes beauty and the importance of symmetry. (A commercial version of this interactive installation has been set up by the cosmetics company L'Oreal in the Millennium Dome where visitors are able to perform a kind of benign Dorian Gray effect: to watch themselves go bald, grow wrinkles and develop sagging skin, and then take home a souvenir postcard – should they so wish.) 'Electronically Yours' also displays a robot face which can interpret facial expressions in the visitor and respond appropriately, emphasising how much of our communication of emotion depends on expression.[41] This robot face reflects recent enormous technological advances in computerised facial reconstruction, especially in the fields of cosmetic surgery and of forensic science. In an article entitled 'The Minotaur Syndrome: Plastic Surgery of the Facial Skeleton', Paolo Morselli describes the surgical procedures performed on a 38 year old man who had serious social problems because of his aggressive and threatening facial appearance that contrasted with his gentle personality. The surgeon coined the term the Minotaur Syndrome to describe the discrepancy between the patient's true personality versus his negative facial appearance:

> *The Minotaur was not a wicked individual, and if it had not been for his facial appearance he would have had a normal life. He was therefore a predestined victim of his facial appearance ... Solving a personal social problem according to the patient's inner self and the public's impression can reduce burdensome psychological problems. These new concepts advance the evolution of the plastic surgery of the facial skeleton.*[42]

In Britain men now account for four out of ten people attending at cosmetic surgery clinics, and in the last ten years there has been a massive increase in the male appearance industry; comparatively few of such operations have been for medical disorders (as above). Copying celebrity looks is a current trend. In October 1998 *The Sunday Times* reported on a number of male makeovers (three men who wanted to look like Michael Masden, Jamie Theakson and Michael Jackson respectively). 'The technology is there', said one. 'Why not use it?'[43]

In forensic science, meanwhile, a new generation of 'cyber-detectives' now use digitised optical imaging to wrap cyberflesh around 3D images of the skull to recreate the face. 'We wouldn't ever suggest that a face we create is a portrait of a person,' says Caroline Wilkinson, a medical artist at the University of Manchester. 'But it should be recognisable as that person.' For such researchers, one of the main imponderables is how to reflect the expression on a person's face. At the 'Human Identification Centre' of the University of Glasgow Peter Vanezis says: 'You have a man's skull, you create a face, but what expression can you give him? If the guy was miserable all his life and you put a smile on his face, no one would recognise him.'[44]

How much of our communication of emotion depends on expression, then? What is the relationship between facial movement and personality? Writing in 1812 in the earliest treatise on the face, Charles Bell notes that 'the thought is to the word as the feeling is to the facial expression'.[45] In *Remarks on the Philosophy of Psychology*, Wittgenstein saw the face as the interlocutor between the self and the world, and facial action and feeling as intimately linked: 'The content of an emotion – here one would imagine something like a picture. The human face might be called such a picture.'[46] In *The Primacy of Perception* Merleau Ponty claims that 'I live in the facial expression of the other, as I feel him living in mine'.[47] For Emmanuel Levinas the uniqueness of the face is that it always remains the face of another, so that it cannot be fully assimilated into oneself. 'The (facial) expression of another does two things: it gives a message I can comprehend, but it also signifies something beyond comprehension – in terms of being able to be explained.'[48] Much of art is trying to understand this, as, in another context, is psychoanalysis. Indeed it is surely surprising that psychoanalysis has not paid more attention to facial perception and recognition to-date. Freud, in his essay on 'The Uncanny', describes an incident on a train journey when, glimpsing a reflection of his own face in a swinging glass door and momentarily failing to recognise it as his own, he feels an intense dislike for the bearded stranger walking towards him. He later wonders if what had happened had been 'a vestigial trace of the archaic reaction which feels the "double" (the mirror reflection) to be something uncanny [*unheimleich*]'.[49] Similar questions were raised by

Jonathan Miller's recent exhibition at the National Gallery in London, 'Mirror Image: Jonathan Miller on Reflection' which centred on the representational ambiguities of the self. 'There is something paradoxical about seeing ourselves from the viewpoint of another person,' Miller notes, '... And yet by the time we are two years old, we scarcely give it another thought: and a significant proportion of human culture is based on the reflected visibility of the personal self .'[50]

Historically and to-date the same issues have fascinated and dogged portraiture: how we perceive ourselves; the relation between artist and sitter, subject and object and the assumptions we make about a person from considering their likeness. Sarah Kofman has compared the good portrait to the uncanny double – to the ghost hovering in a liminal zone, neither alive nor dead, neither absent nor present, staging a duplicitous presence, at once sign of an absence and of an inaccessible other scene, of a beyond. Similarity, she argues, topples all categories of oppositions that distinguish model from copy, the animate from the inanimate and make meaning undecidable.[51] Likewise, Jonathan Cole argues:

> ... to some extent my own face, as seen by me and usually mirror-reversed, may have a more elusive and complex image to me than that of other people. When I think of myself I think of what I have done, or am about to do, and my personal 'face-self' hardly appears. When I think of others, however, faces are very important. Nor can I be sure that the face I see myself, with its mirror-reversal and self regarding look, is similar to the image that our friends and loved ones have. It is very difficult to know, even with photography, and more recently, video.[52]

A Scientific Aesthetic?

While both scientists and artists speak of the deeply embedded role of the face in our perception of the self (and of the face as precursor to cognitive function and even consciousness), the contemporary technological examples offered above reveal continuing contradictions and overlaps in the different epistemological discourses of the face. For, in linguistic terms, what is the register of the language of the face? How is this reflected in the blurring/overlap/exchange between the discourses of art and science? How does clinical discourse combine with descriptive spectacle or perceptual image? How do science and technology appropriate fictional registers and incorporate them? At what point does the intangible/metaphysical/transcendent get translated into a clinical taxonomy of where the image serves to illustrate a kind of knowledge?[53] What is the relationship between external and internal, between a mechanical and a motivational or expressive model of body? How do (can?) we justify the forensic regulation and rendering conscious of a state of interiority in the name of a scientific aesthetic?

If art looked to science in the 19th century, now science and technology look to art: or at least the two are more interdependent, especially in the links between science, art and technology where discursive and disciplinary distinctions are breaking down and/or merging. A Harvard biologist has coined the term 'Consilience' for the presence of science in every sphere of our lives, and in every disciplinary study.[54] As we have seen briefly above, 'consilience' is particularly evident in the mode or presentation of tremendous recent scientific development in the forensic reconstruction of the face. It is also evident in non-forensic medical practice. In 1998 Australian surgeons re-attached a woman's face after it was ripped off when her hair was caught in a machine and British doctors re-built the face of a man so badly injured in a mugging incident that even close friends failed to recognise him. Old family photos were the only guide to restoring his features. The same interdependence of art and science is evident in Oxford's 'Facial Image Archive' where plastic surgeons and artists work alongside the patient to produce a computer generated face that the surgeons then work from during surgery. (Routine face transplants will be possible within five years despite widespread horror at the idea of a possible market for young faces among the ageing rich). Glasgow's Turing Institute meanwhile has developed the 'whole body camera' which is at present used to help surgeons create 'virtual heads' of patients so that they can plan complex facial operations to leave as few scars as possible. It captures a three-dimensional view of a person using a tent-frame system of cameras that feed data into a computer. This then reconstructs the exact contours of the person being photographed. The scientists who developed the system are aware that it would also be ideal for capturing the three dimensional detail of living actors to create computer generated versions which could put dead film stars into new roles. By choosing people who physically resemble dead actors – such as Monroe or Dean – they could in effect resurrect others. In general the only computer-generated film stars to-date have been dinosaurs and space-ships.[55]

The same kind of computer technology (E-FIT and 3-D imaging) is now routinely used in archaeology as well as forensic science to reconstruct the faces of the dead. In March 1998 a forensically reconstructed face of Philip II of Macedonia went on show in Copenhagen. The wax model was created in a Manchester University laboratory after painstaking work carried out by medical experts on fragments of a cremated skeleton from northern Greece. It is one of the most dramatic examples of how modern forensic techniques can help archaeology. It also raises, in a subtly new form, ethical questions about the use of human remains for historical reconstruction. Other recent commissions for the Manchester scientist artists, Richard Neave and John Cragg, include the faces of Robert the Bruce, 'Cheddar Man', a 9,000-year-old skeleton found in Somerset, and Iron Age Briton, two ancient Minoans and three mummies. Neave and Cragg call their faces '3D reports on the skull' to emphasise that these are not portraits, but extrapolations from the evidence of the bones beneath. The impact is remarkable: stripped of artistic subjectivity the faces still draw a strong emotional identification from the viewer.[56]

The main driving force behind all this, most experts agree, is the new technologies, rolling inexorably along. It is now possible to see the face of a 30 week old foetus on the new 3D Kretztechnick Voluson scanner. 'Now many obstetricians take pride in facilitating 'bonding in utero' by acting as showmen who present babies on ultrasound to a delighted audience,' writes Sheila Kitzinger. 'It is the medicalisation of love.'[57] In October 1998 a special issue of *New Scientist* described the emerging technology of germ-line engineering to create designer babies. The vocabulary of one scientist in favour of the new processes is revealing in its synthesis of art and science. Stock claims that:

'Evolution is being superseded by technology – human beings are becoming the objects of conscious *design* ... Genetic engineering offers us the profound power to *sculpt* our children.'[58] The new germ-line engineering is a far remove from the French physiologist Duchenne's collecting heads as they dropped off the guillotine, and applying electric currents to their facial nerves to examine the mechanisms of facial expression in the mid nineteenth century. Such is the speed of scientific development.

With an MRI scan we may be able to image layers of the brain, but scientists tell us that we are no closer to artificial models of consciousness. Some of the current technological elite are less certain, however. In January 1997 Bill Gates is quoted in an interview in *Time* Magazine as saying:

> *'I don't think there is anything unique about human intelligence. All the neurons in the brain that make up perceptions and emotions operate in a binary fashion. We can someday replicate that on a machine. Eventually we'll be able to sequence the human genome and replicate how Nature did intelligence in a carbon-based system.'*[59]

Others, however, disagree. In *Wired Life: Who are We in the Digital Age?* Charles Jonscher argues that the rise of digital technology serves only to underscore the information-processing supremacy of the human mind. He outlines the distinction between raw data and useful knowledge, and makes the point that while computers are very good at handling vast quantities of data, it is still the case that human brains (unlike human brawn) cannot be replaced by machines. He concludes that 'what will retain value in this age of machines that can conjure up and process information in limitless quantities is that which computers cannot produce ... the non-digital quality of human wisdom and human creativity.'[60] So, for many, looking into the mind/self/character can never be an empirical science, and can only be approached obliquely by art. Even then 'There's no art to find the mind's construction in the face', as Shakespeare's Duncan says.[61]

To return to my opening quotation from Liggett. Are we more certain now about 'the questions linking face to character'? Has a further twenty-five years of science and technology helped us circumvent what Liggett calls 'errors of facial understanding'? Are we now certain that the face is precursor to cognitive function, even to consciousness? While, as we have briefly seen above, the nineteenth century was the great age of face-reading, twentieth century psychology has unsettled all that. The relation between facial expression and inner states is increasingly seen as neither a simple nor a translatable matter. According to Maurice Grosser: 'Nobody knows what he looks like. He knows what he feels like inside and that is all. Even when he looks at himself in a mirror, he arranges his face to fit his inside feelings, or else carefully refrains from seeing it'.[62] Thus two people passing a stranger on the street and seeing the 'same' face will draw different conclusions from it. For this reason E.H.Gombrich argues that we shall never know whether we would recognise the 'Mona Lisa' or 'The Laughing Cavalier' if we met them in the flesh. Conversely, as A.S. Byatt points out with respect to her (abstract) portrait by Patrick Heron: 'It is curious how a featureless face can be such a good likeness.' Byatt's remarks recall Gertrude Stein's about her portrait painted by Picasso in 1906. The story of Picasso's 'Gertrude Stein' is well known from Alice B. Toklas's account: in the spring of 1906 Gertrude Stein sat some eighty times for her portrait and yet Picasso was not satisfied; that summer he went to Spain and looked at early Iberian stone sculpture, then, returning to Paris, he completed the portrait, adopting the blocky, stony facial and bold types of Iberian art as the basis of his image of Stein. To objections that the portrait did not look like Stein, Picasso responded, 'That does not make any difference, it will'. And Gertrude Stein herself responded, 'I was, and still am satisfied with my portrait, for me, it is I'. As Richard Brilliant remarks: 'Picasso had so imposed his vision on Stein's portrait that his perceptive 'I' and her receptive 'You' seem to have fused in a proleptic anticipation of the artistic personae they were both to become.'[63]

Scientists wriggle out of this particular dialectical tight corner by articulating a rhetorical face distinction: 'recognition' (or trace presence), as opposed to 'identification' (or the afterglow of human presence). I want to end by sitting fairly and squarely on the fence with Jenny Diski's meditation on face in her novel *The Dream Mistress*. For

Diski's Bella, whose face has been removed by laser surgery, 'face' is neither inner nor outer, subject nor object, it exists somewhere between the living human form, science and art:

Stripped of all its flesh, the exposed muscle, veins, arteries and nerves wove intricate patterns around each other to provide a latticework of scaffolding behind the skin no longer there to conceal them. Delicate blood vessels, veins and arteries, with tiny tributaries running off them like tree roots, snaked between and beneath bundles of striated muscle spanning the space between her temple and jaw bones like rope bridges. ...

But aside from the obvious usefulness of the arrangement, it made a pattern of breathtaking beauty, though not the human beauty which flits around a complete face as the underlying mechanisms create the mobility we generally recognise as beautiful. Without expression, without even a suggestion of its possibility, this face had the cool beauty of architecture or abstract art, and Bella knew it did not tell a truthful tale about how she might have looked in her fully epidermal form. ... But with all the life subtracted, she had acquired a monumental and timeless symmetry, a still perfection of form which almost stopped the heart.[64]

I would like to thank Lisa Lewis for her sharp eye for newspaper cuttings in assembling the material for this paper, and for her criticisms, Mary Ann Kennedy for reproducing the images, and Nick Thompson for providing the inspiration to finish it.

Sordid Sites
The internal organs of a cyborg

Jane Prophet and Sian Hamlett

The Internal Organs of a Cyborg is a CDROM artwork by Jane Prophet which offers cyborg bodies for the voyeuristic gaze. It presents the user with a photostory narrative that combines the visual approaches of the photo-love magazine with the science fiction graphic novel or comic (Fig 1). A text narrative is fused onto scrolling pages of full colour photographs using ready-made images from PhotoDisc's™ stock photography CDROMs. The stock photographs are largely taken from PhotoDisc's™ "Health and Medicine" and "Modern Technologies" archives. The material has been cropped, distorted and montaged to tell the story of virtual lovers from different sides of the tracks whose paths cross in the emergency room.

CDROM Narrative

The photostory describes a young woman from south London. She earns money by participating in drug trials and surgical implant research. She buys additional implants on the black market to satisfy her interest in augmenting her body. Having been discharged from a private clinic following surgery (which involved the insertion of nanotechnological devices) she goes out drinking with friends. Across the bar she sees the repo' man and she flees. He pursues her to take possession of the item that she has bought on credit. During the chase she is shot as the repo' man tries to stop her escaping with the PCID (a microchip which has had her personality downloaded onto it). She is airlifted to an emergency room. Across town the successful director of a chemical bank collapses during the marathon with a heart attack. He is rushed to the same emergency room. Once inside the medical institution their experiences are very different. She has no medical insurance and is therefore offered pain relief but denied expensive life-saving surgery. Meanwhile, in the plush surroundings of a private room, the city gent learns that he needs a heart transplant. The young woman dies as a result of not having surgery and her heart is donated to the head of the chemical bank. But he gets more than he bargains for – inside the donated heart is the PCID implant that she bought on credit and which ultimately cost her her life. The chip contains her downloaded personality and it activates once inside the man's body. His dreams become filled with images of a stranger – a young dark haired woman. When he returns to work after convalescing he starts to receive email from an unknown woman and a love affair begins ...

What follows is a hypothetical Lacanian analysis of the main protagonist in the CDROM. We begin by exploring the Lacanian definition of perversion in relation to the woman. Analysis of this area will discuss issues of voyeurism, the gaze, and identification, the drives and the Mirror Stage. The aim of this article is to question and explore the concept of needs of the human psyche focusing on perversion in the face of possible future transhumanisation.

Voyeurism

The Oxford Dictionary defines a voyeur as 'a person who derives gratification from surreptitiously watching sexual acts or objects; a peeping Tom; a person who takes a morbid interest in sordid sights'. For Lacan, voyeurism is defined through scopophilia which includes both exhibitionism and voyeurism. Scopophilic individuals for Lacanians are classified as clinical perverts. Within this school of thought it is also believed that the structure of perversion is predominately male. Only in exceptional cases could a female be classified as a true pervert. Voyeuristic behaviour however can be experienced and enjoyed by all of us to differing degrees.[1]

The pervert is the person who attempts to take sexual pleasure to the limit in order to achieve the ultimate jouissance. The French term volonte-de-jouissance meaning 'will to enjoy' is often used to describe perversion. This is based on the premise that the individual's enjoyment is derived from the fantasy of a supposed 'other' watching. This is then coupled with the fantasy that they are the instrument of the 'other's' supposed enjoyment. The 'other' is the reflection of the projected self, the ego.[2]

Returning to the CDROM, the female protagonist, Chrissy Hastings, takes the notion of voyeurism to the extreme. She leaves the confines of the 'meat' or flesh body and exists within a microchip that holds her downloaded personality. This chip, complete with the latest high tech cloaking device (which renders it invisible to scanning technologies) has been inserted into her heart prior to her death. Once inside the city gent, her silicon personality is activated with interesting repercussions as 'she' starts to send biochemical messages to the host organism (Fig 2).

Her death is not prevented by the minute technological device, the nano surgeon (a tiny self-assembling machine constructed using nano technology), that she has recently had implanted as part of a medical technology trial. The failure of this technology is essential to the subsequent narrative and gender slippage, and reminds us of the fallibility of the new. The nano surgeon is also a third voyeur in the scenario. This bio medical technology holds Chrissy in its gaze, constantly monitoring her and primed to intercept in any event which it interprets as a medical emergency. The nano surgeon is embodied surveillance, it takes a step further the notion that an awareness of surveillance cameras prevents crime, that the individual monitors his/her own behaviour as a result of being aware of the surveillance camera's gaze. In Chrissy's case the nano surgeon symbolises an internalisation of the panoptican, it resides at the centre of an individual, and is armed with devices to gauge, record and intercept bodily function on an intricate scale.

The Gaze

Voyeurism involves the eyes and the gaze, which for Lacan operate in distinct ways. Looking is from the individual's subjective position, while the gaze is that which looks back at us.[3] We are who we are through identifications with others, such as mother, father, brother or lover. An individual's identity is formed in relation to the 'other'. We look with our eyes at another but the gaze is not owned by us. It becomes the exclusive property of the 'other'; we imagine and fantasize how the 'other' might gaze at us. The gaze becomes the object of the act of looking, the property of the other. It watches the subject from a position that is rooted in the individual's fantasy.

In this narrative we could apply the Lacanian notion of the gaze. By capitalising on the science fiction cliche of being able to transfer human personality onto a microchip, the narrative produces a scenario in which the female character becomes the ultimate unseen watcher. From within the male character she 'watches' his every move. It is absolute; she can 'feel' her subject, monitor his cellular changes and chemical shifts, access his thoughts and brain waves. The microchip could represent the look without the distortion of the eyes. The gaze of the 'other' we could say is being experienced directly from the bodily sensations of the host's body. Our protagonist is literally 'seeing blind'.

The voyeur is focused on sexualising the act of watching the external body. In the CDROM we replace the typical 20:20 vision of the voyeur's gaze with one that takes them inside the body, offering the interior of the 'meat' as the sexualised landscape rather than the exterior. This 'interior' encompasses not only the internal organs but also the innermost thoughts of the character. As the narrative of the CDROM progresses she begins to explore this prime voyeuristic position and gains sexual satisfaction from her internal position. Her 'eyes', supplied via the microchip, are the only area through which she can receive sexual gratification. The 'eyes' are thus erogenised, producing in Lacanian terms the pervert's dream par excellence.

Returning to the CDROM, the narrative further unfolds; she now begins to communicate with the city gent via email. The female character reinstates the danger of discovery as she toys with him (see Fig 3). Users can follow the online love affair between the city gent and his mysterious email correspondent. It emerges that these messages to him are none other than the female 'voice within', sent by the silicon personality of the ghetto girl, Chrissy Hastings. As their love affair develops the city gent embarks on a series of clandestine meetings in the virtual spaces of online chat groups. Here he discovers the joys of Multi User Domain (MUD) sex and text based foreplay.

Who might we ask does the female character fantasize is watching her? What 'other' is she becoming an instrument of enjoyment for? What is the core of her fantasy? Is she trying to find an answer to a question? Or is there a lack of a question? These last two questions are highly significant for Lacanians in the definition of perversion.

Neurosis is characterised by a question that the individual is trying to answer, while perversion is characterised by a lack of a question.[4] Perverts do not question their sexual position in relation to the other, they 'know' they are the means of the 'other's' jouissance.

Drives

The female protagonist in the CDROM has been engaged in surgical intervention and enhancement from an early age. She was perpetually the object of incision, addition and intrusion by medical staff. She is surveyed and invaded by magnetic resonance imaging, x-ray, endoscopy and surgery. Her last operation involved her swallowing a minute robot, a nano surgeon, which was programmed to respond to physiological trauma and to repair internal injury and perform microscopic surgery (Figure 4, page 156).

For Lacan the drives are one of the cores of sexuality. They are based on the premise of never being satisfied. The purpose of the drive is to never reach its goal. Enjoyment is derived from constantly circling and repeating this endless journey.[5] The narrative of the CDROM tells us that she has been involved in drug trials for many years to the point of being addicted. She has numerous bio-technological devices implanted in her body. The character's attempts to exceed and control the limits of her biological boundaries while 'with flesh' become evident as the narrative exposes her use of performance enhancing drugs. We discover areas of the CDROM where we hear her answer phone messages, telling of her experiences of surgery; we read her autopsy report which logs the devices found inside her and lists the scars which mark areas of her body which have been cut open. These are just a few examples of ways in which the body's flesh boundary – skin and muscle – is broken. Audio and animations are used to map a corresponding transgression of gender and social boundaries.

Could it be said that this addiction to technological implants represents an attempt by the drives to realise partial aspects of her desire? If this is true the position that the pervert takes in relation to this journey is very particular. As mentioned at the beginning of the chapter the position of the pervert is to become the instrument of the 'other's' enjoyment as opposed to the will of oneself in the

29

case of the neurotic. Our protagonist would not be trying to find answers to her own desire but would be facilitating the enjoyment of the other. This perverse position means the individual will go to extremes to venture beyond the pleasure principle.[6]

The Viewer

As well as mapping the voyeurism and exhibitionism of fictitious characters, the interactive qualities of the CDROM invite users to satisfy their own voyeuristic urges almost at will. Once they have located links to the email correspondence or other personal effects, they can return to view them whenever they like. In addition the CDROM links to a web site which functions in part as a collection of homepages for the female character. Via the homepages we begin to discover more intimate details about her life, we see the clothes she used to wear, the brands she liked, the places she used to drink, the clubs she used to frequent. We see her as an individual, not as the cyborg without 'meat', which is exemplified in the second interface, made of medical images (see Fig 5). For those of us that have them, our homepages are a means of expressing our innate exhibitionism, and the female character's site is no exception. The web site also plays on the users' voyeuristic drive by offering them an arena in which, not only to watch unseen, but to express their exhibitionism. They can engage in exhibitionism by adding to the email correspondence or by sending in self-portraits and stories, thus displaying themselves to the voyeuristic view of future users.

The Interview

The physical fragmentation, the literal breaking of the body through injury is shown in the CDROM in the operating room scenes. Here the surgeon functions as a kind of 'agent' for the ego, putting the body back together again. In the process of so-called 'invasive' surgery the surgical team breaches the boundary of the patient's body. Part of the research undertaken as part of the production of The Internal Organs of a Cyborg involved audio interviews with surgeons, some of whom were engaged in medical research for implant technologies. One surgeon described a rite of passage which he felt was essential to his exemplary performance as a surgeon. The process of scrubbing up and 'gowning' translated him from the location of consultant (con-versing with the patient in a office and discussing medical procedures in an atmosphere that was often emotionally charged) to the loca-tion of the operating theatre. Once in the theatre he described the necessity of seeing the patient 'as metal, or stone, or wood. The body like a mechanical device that needs repair'. This doctor drew atten-tion to the importance of the surgical sheet as a framing or screening device that obscured the defining features of patients, making it easier for him to see them as 'other'. He described regular moments of slip-page when he looked down in the middle of a mastectomy and recalled a previous conversation in the consulting room, in that moment the patient's body ceased to be meat, stone or wood and they were no longer fragmented. His descriptions of these slippages are reminis-cent of the mirror stage – it is as if the surgeon deliberately employed alienation in order to be able to surgically invade the body, but that his ego struggled against this fragmentation and succeeded in mak-ing him see the patient as whole and complete.[7] When this happened he saw the patient as an individual and he felt he 'was mutilating some poor woman's body'. At these times he literally took a step away from the operating table, for a brief moment, and distanced himself in order to be able to carry out his surgery. The slippage described by the surgeon draws attention to the intimacy between a medical team and a patient and the way in which alienation and frag-mentation can be an essential element of surgical performance. In the CDROM scenes allude to the strangely intimate touching that occurs between doctor and patient, in particular the sanctioned and 'bounded' touching between strangers in the operating theatre, where the surgeon becomes the medical voyeur (see Figure 6, page 156). The surgeon is the watcher that the patient cannot see as they are under anaesthetic (see Figure 7, page 157).

The Fragmented Body

In reference to contemporary western medicine, many argue that technology is harnessed in response to our paranoia and fractured self. Medicalisation could be seen as an attempt to control the frag-mented and decaying body and make it whole again.[8] However the process that the body has to go through in an attempt to find this wholeness can be extremely traumatic for the patient and at times the medical staff.

Imaging technologies take the medical-expert-as-voyeur a step further. Body scanners and heart monitors survey and capture images of the body's most intimate zones and display them for the gaze of the laboratory technicians. These technicians watch the scans, x-rays and microscopic slides unseen by the patient. Numbers classify images of fragments of patients' bodies rather than names, depersonalisation becoming synonymous with patient confidentiality. Patients (especially as described anonymously in medical trials) are reduced to their disease and subsequent physiological and psychological response to drugs and surgery. The medical industry forms a voyeuristic circuit in which the depersonalised body is central. This depersonalised representation of the cyborg body forms the basis of a second scrolling image on the CDROM. A whole female body is laid out, both complete, and sliced, for our perusal (Figure 8, page 157) but less than 5 per cent of it can be seen at any one time as we scroll around it constrained by the parameters of the computer screen (Figure 5, page 156). The CDROM depends on our egocentric need to see the whole body, we are almost guaranteed to want to scroll around and explore the distributed body, represented via scans, x-rays and slices, in an attempt to make the body complete.

The idea of the fragmented body was one of the earliest concepts developed by Lacan in association with the mirror stage. This is the moment in the children's development when they see their body as a whole in the specular register, but their own perception is one of disunity with the image. This time produces great trauma for the child as it is trying to gain mastery over its body. This stage is quickly passed but the trauma of it can re-emerge in later life. The child moves through this period by the formation of the ego. This is achieved by a process of identification with the specular image outside the body. When this happens the individual's ego is formed, based on the alienation process. However, the individual is left 'covering' a lack of completeness in him/herself through the conception of the ego. In other words the ego functions as an agency of deception by telling the individual that he/she is whole and complete. The experience of the 'fragmented body' can result in great anxiety and aggression. For

Lacan the humans continue through their lives constantly oscillating between their image, which is alien to themselves, and their real body which is uncoordinated and in pieces. Many symptoms are in response to the oscillation between these two camps. This sense of fragmentation expresses itself in images of castration, mutilation, dismemberment and combustion of the body, which can be expressed in dreams, one's analysis and in the way we might lead our lives.[9]

In regard to our female protagonist in the CDROM, could we interpret her obsession with technological implants as an attempt to unite the image, which is alienating, and the real body, which is in pieces? If we were to do so, would she still be perverse in some way? If she is, then in what way? In Lacanian analysis the attempt to unite the body in this way could be read as a typical response by the hysteric, whose symptoms are bodily. The hysteric is the classification of neurosis based around the subject's sexual position. The eternal question posed by the hysteric being, 'Am I a man or a woman?' and 'What is a woman?'[10]

Get ALife
Cyberfeminism and the politics of artificial life

Sarah Kember

Creatures

In July 1997 Brighton hosted the fourth European Conference on Artificial Life and Brighton Media Centre held the associated exhibition entitled LikeLife. LikeLife was 'a collection of installations, robots, creatures and artworks inspired by living things'.[1] It included a much publicised Evolved Octopod, a Helpless Robot and a new computer game called Creatures. The Evolved Octopod – 'a large purple creature resembling a Meccano spider'[2] was designed to demonstrate how learning (about movement) evolves in the brain/control system. Artificial evolution 'works best by testing possible brain structures created at random and seeing what works best. The winners are reproduced with some mutations, and the cycle repeats. The results are control systems that work, but we don't know how'.[3] Unfortunately, it didn't work. The so-called Helpless Robot was a cumbersome object with 'minimal visual appeal, yet a strong behavioural dimension' which was learning to 'assess and predict human behaviour'.[4] It had a synthesised voice which it used to ask people to move it where it wanted. The more it got its own way, the more abrupt and bossy it became. Creatures was more impressive altogether.

This game involves breeding and rearing Artificial Life forms called Norns. It is the first commercially available Artificial Life product, and was described by Richard Dawkins as 'the most impressive example of artificial life I have seen'.[5] Artificial Life is an interdisciplinary science which combines biology and computer science and which aims to simulate life as it is, and synthesise life as it could be. Life is defined in terms of the genetic/computer code – or in terms of information. Organisms (both natural and artificial) are regarded as being information processing and replicating systems which evolve, self-organise and are autonomous. In this context, Norns are alive.

Norns inhabit an elaborate virtual world called Albia which has everything necessary to sustain life (food, water, shelter, plants with medicinal properties), to prevent boredom (toys, musical instruments, a film show, a submarine, a tropical island) and to educate (including a virtual computer with a human figure on the screen which acts out simple words like yes, no, left, right and so on). Natural hazards in Albia include poisonous plants and the vicious Grendels; monsters borrowed from *Beowulf* who not only carry diseases and steal food from the Norns, but slap them and make them say 'ow'.

Norns are endearing creatures with their own personalities and very big eyes. From the beginning, the player or 'overseer' of the game is made to feel more like a parent than a genetic engineer, despite the stated aims of the game: 'creating entirely new, and potentially more intelligent life.' Genetic material is provided in the form of six eggs – three male and three female. The player selects an egg, places it in an incubator, waits 30 seconds and is then faced with an apparently wilful creature which must be cared for otherwise it will die. It is quite a responsibility. The thing is no sooner hatched and it's off in a cable car to goodness knows where or trying to play with a Grendel. Norns learn by themselves but can be rewarded or punished with tickles or slaps and, having emerged burbling, can be taught to speak. By selecting the 'view creatures' option from the menu, the player/parent can see from the creature's point of view and can name what the creature is looking at by typing the word on a keyboard. Words appear in speech bubbles on the screen. The creature quickly learns its own name which the player is encouraged to enter on a birth certificate. This displays not only the creature's name, sex and birth date but the name of the 'parent'. It was, admittedly, a moving moment when Bubdis first said 'Sarah'. Norns also speak to each other, breed, grow old and eventually die. They have a life-span of up to fifteen hours. Players are provided with a health kit which displays levels of pain, hunger, exhaustion and boredom; a science kit which includes a scan of brain activity and a performance kit for measuring birth and death rates. When the Norn dies, there is a funeral kit which allows the player to place a photograph on the headstone and write a few words.

This game is not simply about interacting with virtual pets; rather, the player is positioned as the overseer of a process of evolution. The player oversees the evolution of Artificial Life forms with which s/he has a degree of kinship. Norns are like children: a new generation. The introductory tutorial to the game states quite clearly that 'our new-born Norn is alive and like any child she has her own personality'. More than that, as representatives of the Artificial Life project, Norns are the next stage in evolution – a new species. It is clear, within the narrative of the game, that 'we', the players, have responsibility for this new species, but as overseers of the evolutionary process we are evidently not in control of it. We may observe, interact, participate and intervene in the process, but Artificial Life carries forward a firm belief in the sovereign power of evolution. Human agency is at best secondary to the primary force of nature.

The Principles of Artificial Life – and Some Problematics

Artificial Life is at the frontier of contemporary technoscience. Its inception is widely credited to Christopher Langton, Research Professor and director of the Artificial Life programme at the Santa Fe Institute, New Mexico. Other key centres of artificial life research include MIT and the University of Sussex. There is no doubt that this is a male dominated field which nevertheless features some high profile women researchers including Margaret Boden, Professor of Philosophy and Psychology at the University of Sussex, Sherry Turkle, Professor of the Sociology of Science at the Massachusetts Institute of Technology and Pattie Maes, Associate Professor at the MIT Media Laboratory. Boden points to the interdisciplinarity of Artificial Life (ALife) which may be regarded as a development of artificial intelligence and cognitive science. These 'try to model psychology much as work in ALife attempts to illuminate biology'.[6] She argues that the most important concept in ALife, apart from life itself, is self-organisation: 'self-organisation involves the emergence (and maintenance) of order, or complexity, out of an origin that is ordered to a lesser degree.'[7] The emergence of order or complexity is spontaneous or autonomous, 'following from the intrinsic character of the system itself (often, in interaction with the environment) instead of being imposed on the system by some external designer'.[8]

This is where ALife may be seen to be different from, or opposed to AI, where order is imposed 'on general-purpose machines', as it were, from above.[9] Boden elucidates a concept of connectionism which has been taken up in some aspects of contemporary cultural theory.[10] Connectionism is a form of computer modelling which is very different to classical, top-down, AI programming. Connectionism works through a system or network of interconnected units. These systems are referred to as parallel-processing because all the units function simultaneously. Inspired by the function of neurones in the brain, connectionist models are sometimes referred to as neural networks. Connectionist AI models work well with the concept of self-organisation, or bottom-up rather than top-down order and control. What they are modelling, to a certain extent, is the autonomy of the system. Boden states that in top-down processing, the actions or developments of the system are initiated and monitored by the programmer, but 'in bottom-up processing, it is the detailed input of the system which determines what will happen next'.[11]

ALife's focus on the autonomous and indeterminable system correlates with its futuristic element set out by Chris Langton in his seminal paper delivered at a conference in 1987.[12] Once the general principles of life have been isolated, new or alternative life-forms may be allowed to emerge. Boden points out that most ALifers define life in terms of information, and for Langton the principles of this life include self-organisation, self-replication, emergence, evolution and the indeterminable gap between genotype (genetic blueprint) and phenotype (individual). Because life can be defined in informational terms, then emergent life-forms may or may not be carbon based. The one life criterion which appears to divide the ALife research community is the presence of metabolism. Synthetic or computer-generated organisms do not metabolise, but for some researchers[13] this does not prevent them from being alive.

The definition of life as information is a matter of theoretical debate within ALife research. But in the wider field of social and cultural theory it must surely be a highly contentious issue. The definition reduces behaviour, emotion and experience to abstract and computational functions. It effectively eliminates the body, and removes the self or the individual from the social, cultural and political realm.

37

History is subsumed by evolution, and the individual is subsumed by the system. By implication, human agency and responsibility are freely relinquished. Feminist theorists have used the story of Frankenstein as part of a political critique of contemporary medical and reproductive science.[14] Here, science is seen to be fathering itself or creating life autonomously and without the involvement of women – specifically women's bodies. But ALife takes the quest for autonomy a step further. Even Frankenstein had a creator. He was made by a man who was ultimately forced to face his responsibility for his monstrous creation. This was a Romantic allegory of the scientist as God. The Frankensteins of Artificial Life would appear to be godless, without a creator, but perhaps it is science itself – a contemporary science, naturalised by reference to the autonomous, evolving biotechnological system – which is rendered god-like. Science and nature are conflated when life emerges from information processing systems.

While stating that ALife challenges 'our traditional boundaries and categories regarding life itself', Sherry Turkle maintains that unlike AI, the implications of ALife have not been sufficiently disturbing to be played out in popular culture.[15] It is true to say that ALife probably had a more practical than philosophical impact on, for example, film, informing the creation and behaviour of the bats in Batman and the dinosaurs in Jurassic Park.[16] Turkle suggests that 'discomfort arises only when ALife pushes into human-like intelligent behaviour, not ant-like behaviour', and that given this, 'we may only be experiencing a calm before the storm'.[17] But ALife is now pushing into human behaviour with new research projects which attempt to develop artificial societies and artificial cultures.

Nicholas Gessler at UCLA is involved in establishing Artificial Culture (AC) where the goal 'is to create a population of dynamically evolving terrain-based mobile autonomous agents serving as a complex of multiple interacting hypotheses for understanding human cultural behaviour'.[18] The AC software programme will contain virtual food, objects and gender specific 'personoids'. It will also have its own embodied ethnographer and a disembodied, omniscient god.[19] Gessler raises ethical questions as to the status and rights of artificial people, and asks whether human behaviour in both the natural and artificial realms should be judged by the same conventions. Rodney

Brooks at MIT is working on the development of an artificial child. 'Cog' is his situated robot 'baby' equipped with human senses and designed to learn from experience as it grows, but limited to physical rather than cultural interactions with the world around it. Pattie Maes is concerned with ALife applications in the entertainment industry and the development of what she calls autonomous agents. 'Julia' is an autonomous agent who resides in a text-based multi-user simulation environment (MUSE): 'Julia has moods, feelings, and attitudes towards players and a good memory.'[20] She was designed by Michael Mauldin as a "chatterbot".[21] She is able to converse in a "lifelike" manner, where lifelike is defined as 'nonmechanistic, nonpredictable, and spontaneous'.[22] Maes offers an excerpt from a conversation with Julia in which she resists the advances of a character called Space-Ace: 'take a long walk through an unlinked exit, Space-Ace.'[23] Although clearly spirited, it is hard to see from this excerpt where Julia the chatterbot herself makes any advances in the politics of gender identity and relations. Maes also discusses the ALIVE (Artificial Life Interactive Video Environment) project which she helped to develop. This is 'a virtual environment that allows wireless full-body interaction between a human participant and a virtual world inhabited by animated autonomous agents'.[24] These agents are modelled on animal behaviour and include a puppet, a hamster, a predator and a dog. To me, there is some significance in naming virtual characters 'agents' and referring to people as 'users' who have to be trained to interact appropriately with the agents and given simple tasks. A puppet which 'pouts when sent away' and 'giggles when the user touches its belly' infantilises behaviour as well as removing agency from the adult and human world.

Autonomous Agents?

ALife tells the story of evolution from neural networks to autonomous agents who are largely, as yet, not human, not adult and not self-consciously gendered. What these agents are, by definition, is more in control of their destiny than 'god' is, or perhaps, 'we' are. So what is the politics of ALife? To me, a feminist cultural theorist, it could be (amongst other things) a masculine fantasy of autonomy projected onto an imaginary/virtual world of innocent (politically, socially, emotionally) or 'natural' agents. It certainly involves the abdication of (social and political) responsibility to a biotechnological system which is self-organised and evolutionary rather than historical. There is a model of power here, but without responsibility. It is stunningly functionalist and eschews life elements which have historically been feminised: the body and emotions for example. Alison Adam states that ALife research 'involves the study of synthetic systems which are designed to exhibit the characteristics of natural living systems; populations can be modelled over several generations and so they offer the promise of a demonstration of evolutionary biology'.[25] She argues that ALife demonstrates (at best) physical rather than cultural forms of embodiment, and that it is tied to sociobiology which (as a form of biological determinism) is highly politically problematic. Adam makes an important distinction between physical and social situatedness. She argues that the social constructivist position in science studies is concerned only with social situatedness 'as it seems to shy away from dealing with messy bodies, maintaining a masculine, transcendental (albeit not necessarily rationalist) position'.[26] On the other hand, research on AI which attempts to be situated, 'looks at the problem almost exclusively from the physical stance'.[27] Haraway notes that the anthropologist Stefan Helmreich 'correctly insists that the "differently embodied" or materialised entities called information structures, which ALife researchers make and play with, must not be equated with "embodiment" as a point of reference for "locating situated and accountable lived experience"'.[28] At a stage when 'researchers working on the marriage of ALife and virtual reality imagine virtual worlds peopled by virtual living creatures and users' it seems timely to respond to Adam's assessment that 'there is no room for passion, love and emotion in the knowledge created in ALife worlds. A Life's attachment to sociobiological models is based on an essentialist view of human nature and women's nature; where cultural ways of knowing are to be explained and subsumed in deterministic biological models'.[29] This is why a critical study of ALife is urgently needed within both (cyber)feminism and cultural studies.

Cyberfeminism

Despite its problematic core of biological and evolutionary determinism, ALife remains futuristic, unfinished and open to intervention and interpretation. Sadie Plant has interpreted it within the context of her own brand of cyberfeminism. She knits the biological and computer sciences rather seamlessly into an essentialist story about weaving, women and cybernetics in which a feminised technology has become autonomous, self-organised and emergent. The effect of this convergence is anarchic and apocalyptic; it 'looms over the patriarchal present and threatens the end of human history'.[30] For Plant, the metaphor of weaving informs the development of computer software, but also describes the contemporary condition of the net or matrix as a 'web of complexity, weaving itself '.[31] The evolution of the net would appear to be 'bottom-up, piecemeal' and 'self-organising'. It is a network which, 'apart from a quotient of military influence, government censorship, and corporate power, could be seen to be emerging without any centralised control'.[32] Clearly, Plant perceives powerful forces at work to undermine existing authority. Cultures, like cybernetic systems, may be self-organising but intrinsically complex, chaotic and out of control. Man made machines in order to control and subjugate nature (and the feminine), but his creations are returning to haunt him: 'cybernetic systems are fatal to his culture; they invade as a return of the repressed'.[33] For Plant, the invasion is total and the technological revolutions in communications, media and information processing 'have coincided with an unprecedented sense of disorder and unease, not only in societies, states, economies, families, sexes, but also in species, bodies, brains, weather patterns, ecological systems'.[34] In this new, uneasy cybernetic, social and economic climate, women emerge as being fittest and most likely to survive.[35]

Appealing as they may be, uncritical or descriptive uses of scientific paradigms such as this produce a cyberfeminism which is hampered by universalism and essentialism. Faith Wilding and the Critical Art Ensemble recognise that cyberfeminism 'has been largely nomadic, spontaneous, and anarchic' and that this has been a mixed blessing:

> *On the one hand, these qualities have allowed maximum freedom for diverse manifestations, experiments, and the beginnings of various written and artistic genres. On the*

41

other, networks and organisations seem somewhat lacking,
and the theoretical issues of gender regarding the techno-
social are immature relative to their development in spaces
of greater gender equity won through struggle.[36]

Because cyberfeminism is still in its early stages, they argue that 'some feminist strategies and tactics will repeat themselves as women attempt to establish a foothold in a territory traditionally denied to them'. That may include separatism and essentialism which the group regard as developmental stages. The essentialism of VNS matrix is quite radical. Their All New Gen (a computer game with Gen as the heroine) terminates the moral code, sabotages Big Daddy Mainframe ('the omnipresent omniprocessor of a military-industrial complex') and does for Circuit Boy ('a fetishised replicant of the perfect human HeMan, and a dangerous technobimbo') by bonding with the DNA sluts and getting through plenty of G-slime.[37] Recognising the value of this work, Wilding and the CAE maintain that 'the current cyberfeminist mythology will have to fade away'. For them, cyberfeminism 'offers the development of applied, activist theory' which is still being organised around the problems of limited access and the oppression of women. They appear to work with a (Marxist feminist) model of resistance which has been joined, and arguably modified, by models of transformation which have emerged from the theoretically sophisticated cyberfeminisms of, for example, Donna Haraway and Rosi Braidotti.

Towards a Cyberfeminist Figuration of Artificial Life

Haraway and Braidotti work with a concept of figuration which is generally an image or metaphor embodying transformations in the terms of knowledge, power and subjectivity.[38] Figurations form part of an imaginative feminist strategy for intervening in what Haraway has referred to as the 'matrices of domination', including male-stream technoscience and culture. For Braidotti, this strategy is centred on parody. Haraway's infamous parodic cyborg has recently been joined by a triad of figurations: the modest witness, FemaleMan© and OncoMouse™ who inhabit the intersection between science studies, feminism and biotechnology. Modest witness is a figure in science studies, Haraway's subject position and the key to establishing situated knowledge. The modest witness is situated inside 'the net of stories, agencies, and instruments that constitute technoscience'[39] and s/he 'is about telling the truth, giving reliable testimony' while 'eschewing the addictive narcotic of transcendental foundations' in order to 'enable compelling belief and collective action'.[40] S/he works to refigure the subjects, objects and 'communicative commerce' of technoscience, and Haraway declares herself to be 'consumed' by the project of refiguration because she believes that it is central to both technoscience and feminism.[41] She derives the term 'modest witness' from a story about the development of the air-pump[42] in which the modesty of the witness is dependent on his invisibility. This 'self-invisibility is the specifically modern, European, masculine, scientific form of the virtue of modesty'[43] and the one which stakes its claim to truth and objectivity on the basis of disembodiment. This witness appears to be free from his 'biasing embodiment' and 'so he is endowed with the remarkable power to establish the facts'.[44] Haraway seeks to 'queer' rather than oppose the myth of disembodiment and to enable 'a more corporeal, inflected', self-aware and accountable kind of modest witness to emerge within the worlds of technoscience.[45]

It is crucial for Haraway that her modest witness is implicated, that s/he does not seek to stand clear and maintain the dubious distinction between theory and practice, politics and technology, which ultimately reinforces the tradition of invisibility. Rejecting oppositional science studies – and particularly those of Bruno Latour – she argues that 'the point is to make a difference in the world, to cast our lot for some ways of life and not others. To do that, one must be in the action, be finite and dirty, not transcendent and clean'.[46] Preferring Sandra Harding's case for strong objectivity because it 'insists that both the objects and the subjects of knowledge-making practices must be located'[47] she corrects a common misunderstanding about the meaning of location. This is 'not a listing of adjectives or assigning of labels such as race, sex, and class'.[48] It is not as self-evident or transparent as that. Rather, it is the 'fraught play of foreground and background, text and context, that constitutes critical inquiry'[49] and it is partial in as far as it is incomplete and favours some worlds over others. Situated knowledges are then founded on this sense of location, and they are methodologically and epistemologically employed by Haraway in the figure of the modest witness.

Her refiguration of the FemaleMan© is derived from Joanna Russ's 1975 science fiction novel *The Female Man*. It is written as one word in order to highlight its kinship with other sociotechnically manipulated entities such as OncoMouse™. Like the trademark sign on OncoMouse™, Haraway adds the copyright symbol to FemaleMan© because s/he 'lives after the implosion of informatics, biologics, and economics'[50]. S/he transgresses gender and the laws of identity, and is a productive figuration for feminism. Like OncoMouse™ s/he is a natural obscenity who 'might help us rethink the terms and possibilities of a reestablished commons in knowledge and its fruits, more survivable property laws, and an expansive and inclusive technoscientific democracy'.[51] OncoMouse™ is the first ever patented animal and is used in cancer research. The mouse carries a transplanted human tumour-producing gene (oncogene) which produces breast cancer. By carrying a human gene 's/he' is truly our kin and, Haraway argues, suffers so that we (her sisters) can live and can 'inhabit the multibillion dollar quest narrative of the search for the "cure for cancer"'.[52] S/he is a commodity, and like the FemaleMan©, an experiment in transgenesis. Transgenesis was the 'shiny news'[53] of the 1990s. Transgenetic organisms contain genes transplanted from one (plant or animal) species to another and signal a serious challenge to 'the sanctity of life' which, in Western societies is associated with 'racial purity, categories authorised by nature, and the well-defined self'.[54] Transgenesis transgresses the purity of type.[55] In relation to transgenetic experiments and inventions producing entities which are at once material and metaphorical, living and undead, Haraway invokes what is, for me, her most colourfully drained and wonderfully strained metaphor – of the vampire. The essence of the vampire, says Haraway, 'is the pollution of natural kinds.'[56] A 'non-innocent' and ambiguous figure, the vampire has risen up in other cyberfeminist writing,[57] and it signals a desire to identify as and with the other, or 'what is supposed to be decadent and against nature'.[58] Transgressing the categories of race, gender and sexuality 'by illegitimate passages of substance'[59] blood and other bodily fluids which for cyborgs include data or information.

44

Vampire and other such parodic figurations are seriously playful. Through a combination of irony, wit and humour they promise to open up 'spaces where forms of feminist agency can be engendered'.[60] To my mind, figurations could usefully be employed as part of a cyberfeminist intervention in ALife research in art and entertainment. These areas of practice co-exist and are in constant dialogue with theory, and they offer a means of access to a very significant scientific frontier which is largely being forged by an enthusiastic and exclusive posse of cyberspace cowboys. Perhaps in this context Julia would not be designed as a 'chatterbot' for Space-Ace, and perhaps it would be possible to have autonomous agents who are human, grown-up and inclined to learn a different way of being and relating to others in the virtual world. In the grounded and utopian terms of feminist figuration, cyberfeminists could make strategic use of ALife as an unnatural (her rather than his) story of emergence, self-organisation and autonomy.

Conclusion

A cyberfeminist refiguration of the key concepts of Artificial Life would involve a parody of the autonomous, self-organising and emergent agent in as far as this agent is modelled on the myth of the essentially masculine, rational, unitary (humanist) subject. But a cyberfeminist intervention could and should be broader in its aims and begin to incorporate the standpoints of those subjects who have, perhaps, the greatest investment in constructing 'another view of life-as-it-could-be.'[61] Helmreich outlines the masculinist and heterosexual frame of this biologically deterministic discipline which conflates reproduction and sex and renders 'uninteresting those aspects of life that do not have to do with reproduction'.[62] One might also seek to question the racial basis of an evolutionary epistemology which promotes 'the purity of type' (Haraway) or species through selective breeding. So, in addition to strategic refiguration, a cyberfeminist intervention in Artificial Life might entail a hands-on engagement in the construction of 'life-as-it-could-be' in terms which directly challenge or contradict both biological and evolutionary determinism. In another view, life may not be defined in such reductionist and functionalist terms and would not simply be synonymous with information. Other views may reintegrate the body, experience and emotion into an inevitably less computable concept of the organism. It may become possible to refer not exclusively to the evolution of Artificial Life forms, but to their 'materialisation'(Butler) or how they come to matter '(in the sense of both becoming important and becoming embodied)'.[63] The concept of materialisation (rather than natural evolution or social construction) may incorporate a sense of individual and collective responsibility for life-as-it-could-be and might be employed with a feminist methodology and epistemology which produces situated knowledges.

Whose Reality is it Anyway?
A psychoanalytic perspective

Tessa Adams

This chapter will address the problematic of the notion of an 'authentic' reality which is seen to stand in opposition to that which is deemed 'virtual'. The focus will be on the psychoanalytic project with its investment in discerning reality from phantasy: internal from external world, discerning, also, the psychological understanding of that which is self and non-self.

We find that 'reality' in psychoanalytic terms is a problematic which embraces diametrically differing assumptions. We have for example the concept of the 'psychotic' who is characterised as failing to augment fully the necessary symbolic processes which distinguish 'real' from illusionary experience, and by contrast the concept of the 'neurotic' who is characterised as being locked in symbolic representation, to the extent that notions of self are continually in conflict and the decision to act is stifled by ambivalence.

What I am interested in, when addressing certain aspects of the new technologies, in particular the medical technologies and the World Wide Web (WWW), is how to locate the prospect of virtual identities within this framework, and to investigate the language by which psychoanalysis can speak of the fictions that these technologies provide. I see contemporary registers of identity as now paradoxically contrasted between that which is legitimately virtual: computer screen life becoming lived experience, and that which is traditionally lived experience increasingly becoming provisional, as if virtual. Sherry Turkle in her essay, 'Virtuality and its discontents', addresses the complexity of this dialectical opposition. Analysing the language and impact of virtual experience she demonstrates that the 'lived' and the 'virtual' equally mirror the problematic of authenticity. We are experiencing a 'liminal moment', she tells us, 'as we stand on the boundary of the real and the virtual', a moment of passage when new cultural symbols and meanings can emerge.[1]

But what of the subject who fears this moment and all that new technological systems provide? I am thinking here of the large proportion of individuals who, in reporting symptoms of psychosis, cite some aspect of communication which they cannot control. The bulk of which is seen to spring unbidden from the twentieth century technological advancements such as the radio, television, telephone, electric light, etc. Of course, threat from that which is mysterious and 'out there' has not been restricted to this century. We could probably trace equivalent psychotic triggers over the ages, but the terror of the transmitted message currently occupying so many cases of paranoid-schizoid anxiety cannot be ignored.[2]

For many individuals in states of psychosis, the threatening communication is seen to come through a human agent who is conceived as having powerful systems at his/her disposal. A sort of 'Big Brother' syndrome which is maintained by the phantasy[3] that all those who question the delusionary experience are complicit and under the influence of the 'force' that the sufferer believes is 'taking over'. In other words, so insistent are the dominant phantasies in psychosis that all communication can be interpreted as an aspect of the dilemma and therefore potentially malevolent. The case of the man who was convinced that MI5[4] communicated directly through his penis is one such example. So complete was his terror of invasion that psychiatrists' attempts to convince otherwise were met with equal fear, for 'doctors' were seen to be collaborators with the aggressor. This was someone who would prefer to account for his vulnerability as mysterious, forbidding and beyond his direction. It is significant that the onset of this phantasy occurred at the time when he experienced seeing his father hospitalised, in intensive care, coupled to an elaborate system of drips and lines; a technological intervention which failed to save his father's life. Arguably, this shock of the invaded dying body, to which the medical technologists administered, opened up the delusionary prospect of MI5 determining a 'direct line' to the most vulnerable part of this patient's anatomy.

There is also the patient (classified as schizophrenic) who claims that the dentist has stolen his body and who has been attempting for many years to convince all those around him to repatriate him with the body that appears to have been taken. He is certain that the one he has now is not his. It is as if the impact of the technological intervention of the dentist's chair has been so devastating that his entire identity has been wiped out. Struggling with the 'wrong body', this psychic tragedy remains vivid to him but his protests are now subdued; time has taught him that the psychiatric staff and his family have no power to retrieve the body that he mourns as lost. For these patients, and many others, psychotic anxiety is located in relation to a body that seems to be no longer intact or under their control. Such symptoms are commonly classified as a thought disorder and attempts to reinstate stability through psychiatric intervention can often be ineffective. A rift has occurred, external and internal have become confused, rendering the boundary and ownership of the body constantly in question.

Of course psychotic anxiety centred on the fear of corporeal instability is not new. I am not arguing that developments in technology are responsible for the many volatile expressions of psychosis, yet it does seem probable that for certain individuals the increasing medical command of the body, both through literal and digitised investigation, verifies the prospect of the body's fragmentation. In their book *Body/Politics*, Jacobus, Keller and Shuttleworth discuss the colonisation of women's bodies as 'masculine' science' determines its historical objective. I suggest that medical discourse, with its body narratives and folio of imaging techniques, provides a fertile forum for symptomatic response.[5]

Let us reflect on how tenuous our own body stability might be. Surely the patients presented above represent the conflict that we all touch upon each time we are subject to the dentist's or surgeon's excavations, disintegrating our illusion of body wholeness. The cavity of our internal vulnerability reminds us that we can become subject to increasingly elaborate systems of somatic plunder as medicine genuflects to the icon of peak-health. As we spit our debris from the dentist's chair, touch the stripe from the surgeon's wound, or seek our body part in the scanned image, it is then that we know that we have relinquished dominion. In the most part technological intervention is feared and the greater the elaboration of the intervention for cure the greater the anxiety of disempowerment.

In contemplating the impact of new technologies, in particular those systems that increase our exposure to the fabric and function of our physiology, several questions come to mind. First, will our (Eurocentric) illusion of corporeal autonomy[6] be increasingly challenged by contemporary medical preoccupations in which the codifying of the body's intricate systems predominates, particularly through interior investigation and visual representation? Second, with the rush of technological innovation that modern medicine fosters, in experiencing devices which penetrate the body's boundaries illuminating and quantifying our very matter, can we conceive an 'inner world' that would remain intact? Here 'inner world' is that 'concept of self' specific to certain psychoanalytic practice which privileges the notion of 'authentic identity' as a key to psychological well-being. Third, does the prevalent media reportage of the body's silent dramas, intentionally coercing us to form a new intimacy of 'seeing inside', revision our physiological and psychological identity?

Sarah Kember has approached certain of these questions in her book *Virtual Anxiety* in which she argues that 'science and technology are fully cultural and ideological processes' and that our response to them is driven by both unconscious and conscious investment. She rejects the view that we can stand apart from the machines that we have constructed, claiming that we should be 'politically and socially responsible for the futures that we create with or without' the technology to which we have become psychologically bound.[7] She asks 'what kind of subjects are we becoming, and what kinds of subject do we want to be?'[8]

Consider the impact of the current medical imaging technologies which profile aspects of pregnancy as one example. The development in digital scanning has radically shifted the idealisation that maternal experience can remain in the private sphere of mystery to that of public display, and thereby public ownership. The visible life of the foetus projected on the screen addresses all who are nominated to evaluate its survival: all those who interpret its status. And since inter-utero life, in the wake of this technological advancement, enters the arena of judgement, the neonate is signified as 'other' through its digital representation.[9] That is to say it belongs 'out there', classified either to be accepted unconditionally or to be rejected. Within this shift from

the ideal of untroubled (private) pregnant anticipation to the elaboration of public appraisal, motherhood is tested, since it is quantified in
terms of contemporary analyses of sufficiency. Carol Stabile explains
this 'maternal' crisis as an 'ideological transformation of the female
body from a benevolent, maternal environment into an inhospitable
waste land, at war with the "innocent person" within'. Her question is
whether 'this historically unprecedented' split between the woman and
foetus is socially progressive or regressive?[10] She speaks of the 'enormous repressive impact' of the visual technologies used to 'isolate the
embryo as astronought, extraterrestrial, or aquatic entity in the legal
and medical management of women's bodies'. Yet she does not ignore
the aspect of deconstruction in this project of digitised foetal autonomy,
in casting that which has been the reactionary subject of 'feminine
nature' to the winds of medical possession.[11]

Death

Although a different aspect of technological advancement, the issue
of private and public ownership also is at stake in the case of medical death. As we know medical technology is employed to defeat
mortality by systems of surrogacy in which resuscitation of the body
is achieved through life-support. It is an enterprise from which a new
category of individual has emerged, namely, the 'brain dead', whose
existence is tenuously leased to the machine environment. Is this not
a realm of splitting which concretises the psychotic conviction that
the mind and body can operate discretely? How much less bizarre is
the account of the dentist stealing the body in this context, or the
characteristic psychotic fear that the mind can be taken over by some
kind of legitimised authority.

With medical advancements so highly tuned that the body can be
sustained in death what does it now mean to die? The predominant
rhetoric is one of failure with the paradoxical question – was it the
system that failed the patient or did the patient fail the system? This
is an example where technology and biology combine to question our
human status, exemplifying Turkle's argument that the 'traditional
distance between people and machines has become harder to maintain.' Her claim is that we are so enmeshed in our new technological
relationships that we are 'obliged to ask to what extent we ourselves

have become cyborgs, transgressive mixtures of biology, technology and code?'[12] It is this confusion between biology and technology that has engendered a movement to remove the dying body from intensive care. Yet, bidding for a technological free zone, those who seek the subject's right to die have introduced a further paradox – the 'living will' as a means by which to reinstate the subject's control. Ironically, though, medical science is seeming to 'have the last word', for with the advent of cloning mortality is finally challenged by the prospect of our body's replication and continual regeneration.[13] Here the psychotic phantasy of 'body theft' is realised, but not the cumbersome fleshy frame taken from the dentist chair, simply a surreptitious stealing of the co-ordinates of the substance of life itself.

Seeing Inside

Despite the elaboration of the technological culture for cure, with its imperative to ensure longevity and the ethical questions that medical imaging raises, the ambition entirely to decode the body's information appears to be limitless.[14]

Psychoanalytically the dynamic of this ambition can be interpreted as generating from our original infant illusion that we created ourselves.[15] Yet the greater the complexity revealed, the greater the risk of psychological dissociation and dislocation. Amidst this explosion of discrete data that locks us to a fictional notion of familiarisation with our body processes (through media analysis of our body functions) how fragile is the consolation that this corporeal intimacy offers. Is there any real comfort in colluding with the bio-chemist's wish to display the instability of our cells? How does it feel to be confronted with the animated spectacle of our matter or the demonising of certain of our genes, and the deconstruction of the many other mysteries of the body's structures; let alone the prospect of reincarnation from our cloned genesis? Surely, these techno-maps of our biological commonality which (through scopic lenses) beautifies our digestive juices, lymph nodes, urinary tract and coursing blood, mark the suffering of our powerlessness enhanced by the internal exposure. An exposure for public broadcast as we are invited to witness the prising of a diseased hip from its socket, the bloody spectre of the triple heart bypass or the

path of cancer's stealthy victory. Here is the blood sport of life codified, mechanised and aesthetisised and, struggling to make sense of
a mass of body detail (partially informing us), we seek to locate those
elements that would constitute 'self'.

What has happened is that the anatomical drawings and investiga-
tions of cadavers of earlier centuries have now come of age; no longer
simply to service medical enquiry, but to engage us in a three dimensional manual of life. Arguably, for some this 'manual' can offer
reassurance, in that the subject can become a medical expert of his/
her own condition. Or it might be that the intimacy of 'seeing inside'
mitigates the primacy of the 'master discourse' when the actual personal body is revealed. Certain body artists, Orlan for example, have
redefined their relationship with their body's interiority by a radical
engagement with its material structure.[16]

Hiding Place?

Notwithstanding the exceptions above, I suggest that the means by
which we secure identity, in the face of the increasing complications
of the biological map, is to move entirely away from the vicissitudes of
the corporeal. That is to say, running parallel to the medical project
of technological saturation, there exists the oppositional technological advancement that allows us to 'get out of the meat'[17]: this is the
culture of simulation, the realm of our virtual presence and all that
defends us against the instability of our embodiment.

It is significant that contemporary technology encultures us to define ourselves through these contrasting systems, the first through
biological examination, the second through conceptual configuration.
Psychologically it could be argued that the impact of so much techno/
bio investigation has engendered a split state in which our desire to
reinstate our omnipotence is effected through the elaboration of our
bodiless potential. This is to suggest that we have designed an arena
in which our physiological vulnerability is mitigated by a realm of
experience that eschews all prospect of corporeal intimacy, creating
in its place a non-biodegradable environment in which the body's absence is permanently sanctioned.

Codified as the WWW this environment offers identities that are
paradoxically enduring and interchangeable; gendered and

53

ungendered from which vantage point a mercurial present is forever animated, forever productive and forever young. Instead of pondering our body's struggle for homeostasis, in a leap of digital extravagance we can join a Net-work of hide and seek thereby to barter our corporeality for a fictional universe to become transmutable, transgendered players. Entering these alliances, promiscuous, energetic and ageless, we are free from the constraint of mortality. As alternative athletes we saunter in a stadium where no hamstring is pulled, weight ratio questioned or muscle fatigued. We construct an ideational body which develops and propagates with more and more elaborate systems of codified communication, a self that is both commanding in its signification yet continuously under revision.

Thus it appears that we are caught between two technological stools. On the one hand we have at our disposal a vast matrix of data disturbing the last mysteries of our body functions; an electronic system of investigation which emblazons before us the complex wilfulness of our substance. On the other hand we employ that same digital edifice to furnish a safer and more enduring fictional universe, weaving a web to cast away all fear of our temporality. But what is the status of this Utopian liberation from the corporeal as we court the virtual body of cyberspace?[18]

What about ALife?
Perhaps it is the final irony that, in seeking to relinquish the body, we have created the crucible of artificial life (ALife).[19] What does it mean to teach a mechanism of absent flesh to evolve itself? Do these parodies of life, that this demanding technology structures, restore our omnipotence (for here is a world that we can finally make happen) or are they simply icons for identification? In the face of increasing interest in technogenetic generation it would seem that we have become both attendants and witnesses and, as ALife models itself from the template of our humanity, the question of ethics arises. Kember is concerned about our responsibility towards the affect of the 'creatures' that we create.[20] Clearly our emotions are involved, for the biomorph is both 'of us' and 'of itself', calling into question the co-ordinates of what is to be an 'organism'. Richard Dawkins, the biologist, when developing his

54

simple structures that mutated to form spider-like insects, exemplifies this point. Recalling the moment of their genesis, he reports:

> *I still cannot conceal from you my feeling of exultation as I first watched those exquisite creatures emerging before my eyes. I distinctly heard the triumphal opening chords of* Also sprach Zarathustra *(the "2001 theme") in my mind. I couldn't eat, and that night "my" insects swarmed behind my eyelids as I tried to sleep.*[21]

How can we estimate the psychological impact of this split between our degradable biology and indestructible technology in the context of our primary fear of the body's fragmentation? In running over clouds without getting wet, is the potential of psychotic anxiety mollified by entering an environment in which the exchange of information constitutes dominion? That domain in which digital supremacy furnishes the illusion that human dissolution can be kept at bay. Arguably our determination to release ourselves from the impediment of our mortality has engendered a territory in which the psychotic phantasy of a malevolent automated authority is anticipated. The prospect that technological systems might gain the potential to outreach and outwit us is at least conceivable. As Turkle asks, what would be the outcome if the autonomous 'creatures' of ALife were imbued with a superior capacity of Artificial Intelligence?[22] In this light the common psychotic fear of being taken over by technology would surely seem to be rational anxiety instead of pathological delusion?

What Are We Doing?

What is this fascination with imitative life which drives us to reproduce ourselves in that paradaic digital womb? While we construct technology's offspring so that they become the performing jesters of our personal kingdom, where do we locate ourselves? How do we reconcile the cost of spawning life's substitutes in the wake of starvation's cry? How do we assimilate the bid for technological resources designed to nurture our virtual issue in the face of the actual suffering child: that child whose playless world (of famine, war and disease) cannot meet their needs? The paradox is obvious when unsolvable human distress stands in opposition to the more acceptable challenge of the creation of artificial life. Psychologically this anomaly can be interpreted in terms of the seductive aspect of the computer screen which offers a reflection for our own narcissistic ambition. In other words, psychodynamically, the screen becomes that mirror in which the infant first beheld her/himself; omnipotent in action, with every gesture gratified by simultaneous reflection. From this position the response of each reflected image creates the illusion that the world is subject to sublime control. This dynamic of narcissistic attachment is exemplified by a comment made by an adolescent schoolboy. In explaining how his four (screen) personalities function he demonstrates that real and virtual experience have become narcissistically inter-twined: 'I split my mind ... when I go from window to window ... RL [real life] is just one more window, it's not usually my best one.'[23]

Arguably, in Lacanian terms, in this system lack is denied, the objective being to mitigate the absence of the 'real'. For the substantive dynamic of each act in web-space, through the material of its technology, succeeds by privileging symbolic coherence.[24] As Narcissus gazing at our reflection, ignoring Echo (the unrepresented – the hungry and dispossessed) calling in vain for our glance of responsibility, we remain infinitely self-generative. In this pool of digital reflection, autonomy is forever yielding, a site of self-serving in which conflict is dissipated by the never ending supply of World-Wide participants who are prepared to collude with our narcissistic ambition. That which is virtual, with a paradoxical shift, can be seen to both challenge and resort to reason. In this environment the exponential universe of virtual identities is the sublimated one. Who are we now as the creator of the 'Anti-Body'? Is it the way to further identify ourselves or are we simply to become puppeteers who shroud their corporeality permanently from view?

This question is one that I cannot answer. Perhaps this chapter has overly stressed the contradictory aspects of certain of the new technologies, polarising the split between biological necessity and conceptual extravagance. What is missing in this exegesis are the representations of our humanity which break through the screen of our technological superiority. For example, there is an aspect of the 'chat rooms' that offers the prospect of 'messing about'. A domain of activity that is bodiless, spontaneous and versatile, but includes the potential for reflection, serving as an antidote to the imperative to be autonomous and productive.[25] We can see this phenomenon operating in terms of the Winnicottian principle of the 'potential space to play', so essential for the maintenance of self discovery.[26] A transitional realm, neither within or without, that is designed to mitigate the anxiety of solitude. Alternatively we can draw a parallel in Kristevian terms, by identifying the creative-play of chat rooms as the site of the semiotic[27] designed to undermine the relentless bid for symbolic coherence. In other words, this presence of a subverted discourse could be interpreted as the eruption of 'maternal signification' through the screen of the patriarchal order?

Acting Out[28]

Finally there is the case of the 'Hacker'. The saboteur whose capacity to cavort amongst the digital air waves threatens the dedicated participant. This is an individual who, in Jungian terms, personifies the archetypal trickster whose job it is to challenge the stability of the status-quo.[29] The subversive whose innovative capacities are not used to further the development of new initiatives, rather their expertise, intentionally sinister, subscribes to the illusion that effective systems of communication can be disassembled. There are those who spawn their 'viruses' to course through the networks of institutions and, like arsonists, delight in witnessing the effects of destabilisation. These interventionist acts, although often economically targeted, expose our psychological dependency: a dependency that has been encultured by our developing integration with the information that enshrines us.

Further, the hacker in evoking the spectre of disease reinforces our corporeality. In the face of continual fear of digital viral invasion we seek immunity as if it is our own body that is at stake. But what has been significantly exposed from this aspect of the tricksters 'bag' is the tenuous nature of our machine alliance. For we cannot be sure to fully protect the machine from 'infections' any more than we can be sure to protect ourselves from the biological equivalent. Andrew Ross, in his essay 'Hacking Away at the Counterculture' makes this clear. He states that a 'climate of suspicion' now exists between users that is difficult to overcome, citing the exaggerated 'health warnings', framed in AIDS rhetoric, that have charged the computer virus debate with what he terms 'hysteria'.[30] What is interesting in respect of the anxiety surrounding computer 'hygiene' is that it is as if the paranoid anxiety of the 'psychotic' has been reversed. The focus here is on fearing a malevolent human intervention on the 'brain' of the machine'. This is an ironic mirrored contrast to the 'psychotics' delusion of technology's threatening attacks. A further irony is that in each situation there is the same dread, namely, the fragmentation of 'reason'.

Then there is the 'specialist'. The hacker whose unlawful entrance can threaten governments and institutions. Like Oedipus, unrecognised, slaying the father at the cross-roads (breaking the code) the forbidden bed is entered and all that has been concealed is exposed. Here is the omnipotence of the 'imaginary' challenging the Symbolic Order and chaos ensues. These subversive acts of system entry bring the Law into question primarily because the real body is missing. Is it trespass, theft, or simply virtual crime? Bill Landreth's response (as a celebrated boy hacker) to the judge at his trial emphasises this point.

> *I am pretty much a normal American teenager. I don't drink, smoke or take drugs. I don't steal, assault people, or vandalise property. The only way in which I am really different from most people is in my fascination with the ways and means of learning about computers that don't belong to me.*[31]

As you can see, 'real' stealing and vandalising property is regarded as a criminal contrast to the innocence of his 'virtual' breaking and entering. But it is a question of perspective, for the judge sentenced him to three years in prison – no doubt this 'RL window was not his best one'!

58

Section 2
Outertextualities
Internal Sites and External Experiences

Wired for Violence

Sherry Millner

> *The violence inherent in space enters into conflict with knowledge, which is equally inherent in that space. Power – which is to say violence – divides, then keeps what it has divided in a state of separation; inversely, it reunites – yet keeps whatever it wants in a state of confusion. Thus knowledge reposes on the effects of power and treats them as 'real'; in other words, it endorses them exactly as they are .... In the dominated sphere, constraints and violence are encountered at every turn: they are everywhere.*
>
> *Henri Lefebvre,* The Production of Space.

In 1954, Maxwell Anderson's play 'The Bad Seed' became a solid hit on Broadway, thrilling audiences night after night with its daring depiction of a cute little blonde girl in pigtails and polka-dot dress who was also a born killer. At the time, the critic Eric Bentley, who was minimally impressed by the spectacle, drily commented, 'The homicidal child is a traditional source of quiet fun.' In the Fifties, the child who killed was much more a theatrical invention, a representation of adult anxieties about child-rearing deliriously displaced to the unknown realm of the genetic, than it was perceived to be a reality. A generation later, the adult fantasy of the murderous child resurfaced in 'The Omen' (1976) and its sequel, 'Damien: Omen II' (1978), giddily lurid occult films which detail the problems ordinary parents have in raising a youngster who also happens to be the Anti-Christ. Satan's spawn proved to be quite a handful – truly, as they say, beyond redemption. These films' grotesque but canny appeal to the irrational and demonic underlined how little public credibility there was to the notion of the underage murderer in the Seventies. On the other hand, by 1979, in San Diego, sixteen-year-old Brenda Spencer opened fire on the elementary school across from her home, using a .22 caliber rifle, killing two, wounding nine. Her explanation for her actions still has an off-hand resonance more chilling than mere demonism. Talking on the phone to reporters while holding off the police, Spencer said, 'I don't like Mondays.' Such casual sound bite nihilism got her 25 years-to-life, but which of us has not felt something akin to a murderous rage upon waking on a blue Monday to face another week's submission to routine?

Slow dissolve to a generation later, and what still seemed like almost undiluted adult fantasy had migrated off the silver screen and onto the front page. Between February 1996 and June 1998, at least five lethal shooting sprees occurred on school grounds in the U.S. – all of them perpetrated by suburban white male students between the ages of 11–16. In each case, the media solemnly greeted these schoolyard massacres with the following question: How could this happen? Or alternatively: What makes these boys kill? What made Luke Woodham, age 16, kill his mom and then two students at his high school in Pearl, Mississippi? And what pushed Michael Carneal, age 14, to kill three students and wound five, in West Paducah, Kentucky? And what inspired Andrew Golden, age 11, and Mitchell Johnson, age 13, to open fire outside Westside Elementary School, in Jonesboro, Arkansas, killing five and wounding ten?

This outcry was inevitable and perfectly understandable, but its framing was almost completely disingenuous. Middle-class suburban white boys personified America's 'Troubled Youth'. Suffering from self-induced blindness, the media scared itself silly seeing this class-specific, race-specific, gender-specific, exurban-specific gallery of killer kids as emblematic of all its youth, a category error so egregious that it speaks to the media's own belief in a hegemonic standard of normalcy that it was instrumental in creating and still perpetuates. This misreading begot a strident alarm that made it abundantly clear exactly which 'Troubled Youth' 'America' is supposed to regard as its own, a view heretofore incompatible with mass murder, somehow or other.

Nevertheless, terrible crimes of such an exceptional nature inevitably evoke desperate attempts to identify a root cause, preferably a singular root cause, even though it's in the nature of roots to proliferate madly. We want so frantically to know why – so that we can stop such events from ever occurring again. And we would prefer to have an easy answer – like nature (the Bad Seed phenomenon: born-to-be-bad babies imprinted with satanic DNA) or if nature's found wanting, nurture – in short, bad parenting or its sinister variant, child abuse. We are all too well aware of the legions of inept or damaging parents out there. Walk into any supermarket in the land and you see the poor grown-up dummies screaming at their toddlers, whopping their two-year-olds across the butt. What's the classic line? 'Stop that damn cryin'' or I'll give you somethin' to cry about.' And then they do. Since as high a percentage of American parents believe in the efficacy of physical punishment as believe in the existence of Satan, they think nothing of beating on their own little devils. However, if bad parenting were the answer we wouldn't be talking about a mere handful of arresting crimes. We'd be talking about shoot-em-ups on every schoolyard.

Interestingly enough, following the Columbine High School massacre in Littleton, Colorado, on April 20, 1999, in which fifteen people were killed, including the double suicide of the young killers, precisely that dire possibility was forced upon the media. Within a week something like 20 per cent of U.S. high schools had been evacuated after bomb threats. Suburban white boys all over the country, either inspired by Littleton or found to be planning their own mayhem, were arrested. And then, on May 20, in Conyers, Georgia, fifteen-year-old T.J. Solomon, a church-going Boy Scout, fired on fellow students at his high-school, wounding six. That same night, the mounting pressure of such events forced the U.S. Senate, which has lounged in the well-lined pockets of the extremely right-wing National Rifle Association for fifty years, to pass what the media hastily and inaccurately called a gun-control measure. This legislation was supposed to help protect children from easy access to purchasing and using guns. However, it was, more pertinently and problematically, encased within a strong juvenile crime bill, a battery of provisions to punish wayward children, age 14 and up, more severely by treating them as adult felons. Exemplary patriarchal reasoning at work: if beating up on the kids doesn't do the trick at age two, then lock 'em up at fourteen and throw away the key.

So what had seemed a rare, if disturbing, phenomenon perpetrated by a few misfits now seemed a breathtaking alignment of a series of forces suddenly surfacing all at once that called into question the socialisation of suburban middle-class white boys all across the country. What had seemed before Columbine like scary but relatively anomalous ripples in the good-and-plenty culture of suburban male youth had now definitively cracked to reveal an abyss. In the U.S. nearly every newsworthy social phenomenon quickly morphs into a TV movie, exploiting, if not exploring, popular fantasy. But if an event is creepy enough even TV will give it a pass until the heat's off. We've not been treated to any schoolboy schoolyard massacre movies, as yet. The blood's still wet on the teeter-totter. What happened instead was that certain TV series and specials were delayed. For instance, the season finale of the Fox Network series, 'Buffy the Vampire Slayer', apparently featuring an exciting bloodbath at Buffy's high school, was put off until the fall season. Fictional depictions of teen violence were temporarily seen as tasteless. Meanwhile, on what passes for the public sphere in the U.S., the three all-news networks (CNN, Fox News, MSNBC), produced their own non-stop orgy of hand wringing coverage, often cutting between battle scenes in U.S. high schools and in Kosovo. Fantasy, in other words, was regarded as much more dangerous than reality. And with good reason: fantasy had, for the moment, taken over the field of action. What was happening revealed that the normative standard of good solid middle-class Christian suburban white male youth, the heretofore unassailed 'Gibraltar' of teendom, had become such an oppressive fantasy that even some of its own presumed beneficiaries were willing to tear it to bits.

But who were the cracked kids who cracked 'the Rock'? The media rapidly assembled profiles that taken together produce a fair verdict. What seems to have happened is, to quote the title of yet another film, the 'Revenge of the Nerds'. High school is regarded alternately and with far too little metaphoric exaggeration as jail or war by many if not most students. Certain of the losers in the suffocating class structure of high-school life – in which the top dogs are still the jocks and the cheerleaders – had reached the breaking point, armed themselves and become teen terrorists. Since they were losers nobody had paid them much attention until it was too late. Riven by *ressentiment* largely due to the hostile terrain of the 'good' schools they attended – invariably schools that smugly boasted 'it can't happen here' before it happened here – such severely depressed, often medicated kids were vulnerable to acting out inchoate fantasies of revenge.

Considering the relative coherence of this assembly, the individual psyches of these kids is less significant than the enabling factors, the triggers, that make this phenomenon possible and even perhaps inevitable at this moment in this culture. Given that such high schools have existed since the white flight to suburbia in the Fifties, what is distinctively new or different about the culture now that enables such an unprecedented assault? The massacres represent an extreme response to certain conditions which just about every student feels to some degree (to judge by the nationwide response) but only the most vulnerable kids act upon. Every TV psychology expert noted that levels of depression among teenagers are at an all-time high. The interesting question then is why violence seems the only way out to these kids to the degree that they imagine themselves as embattled terrorists and their own schools as battlefields and then act upon that fantasised perception.

A suggestive book, *On Killing,* by Lt. Col. Dave Grossman, makes a strong case for how extraordinarily difficult it has been in each war until the present era to successfully train soldiers to kill, to actually fire their weapons during wartime. Grossman documents how extremely resistant ordinary human beings are to shooting each other, even under what would seem to be the most highly motivated conditions of the battlefield. He equally suggestively maintains that the same conditioning processes which were at last so chillingly successful during the Vietnam War are now at work among civilians – in effect, potentially creating a new generation of teenage Rambos. Grossman's book documents the shift since World War II, in which, even under direct attack, no more than 15 per cent of combat infantry were willing to fire their rifles, to the Vietnam War, in which 95 per cent of the mostly teenage soldiers fired their weapons in battle. He attributes this startling change to a new technology of conditioning developed to overcome young men's natural resistance to killing. To quote Grossman:

> *The three major psychological processes at work in ena-*
> *bling violence are classical conditioning (à la Pavlov's*
> *dog), operant conditioning (à la B.F. Skinner's rats) and*
> *the observation and imitation of vicarious role models in*
> *social learning. In a kind of reverse 'Clockwork Orange'*
> *classical conditioning process, adolescents in movie thea-*
> *tres across the nation, and watching television at home,*
> *are seeing the detailed, horrible suffering and killing of*
> *human beings, and they are learning to associate this kill-*
> *ing and suffering with entertainment, pleasure, their*
> *favourite soft drink, their favourite candy bar, and the*
> *close, intimate contact of their date .... Operant condition-*
> *ing fire ranges with pop-up targets and immediate*
> *feedback, just like those used to train soldiers in modern*
> *armies, are found in the interactive video games that our*
> *children play today. But whereas the adolescent Vietnam*
> *vet had stimulus discriminators built in to ensure that he*
> *only fired under authority, the adolescents who play these*
> *video games have no such safeguard built into their con-*
> *ditioning. And finally, social learning is being used as*
> *children learn to observe and imitate a whole new realm*
> *of dynamic vicarious role models, ... often portrayed as a*
> *murderous, unstable vigilante who operates outside the*
> *law.*[1]

I don't know about you but this argument struck me as almost ab-
surdly simplistic until I realised that was the strongest thing in its
favour. The kinds of primitive conditioning Grossman describes work
at a basic level on species bred for aggression – on dogs, rats, adolescent soldiers, and now adolescent schoolboys, hungry for action. Not
very many schoolboys, just a handful – a receptive handful in every
suburban high school, that is, ticking away like nearly human time
bombs. Less vulnerable kids presumably use the vastly pervasive technological armament of training for violence in civilian life to get off
on merely imagining what they would like to enact rather than en-
acting it. Certainly that fantasised form of aggression is enough for
most of us. But it does show that there's still something of a

shortfall in the perfection of training, since most middle-class suburban white boys exposed to similar conditioning fail to become killers.

Many parents have thus far approved of this technological apparatus – particularly the video games. They keep the boys quiet and off the streets – usually. In the case of both action films and video games, computer-based special effects and graphics are keyed to enhance what is invariably called throughout the industry and by audiences alike, 'realism'. The thrill of such so-called realistic special effects (an effect of the blockbuster phenomenon ushered in by 'Jaws' and then 'Star Wars', with the George Lucas company Industrial Light and Magic as the presiding R&D maestro), is that every new technological advance makes the fantasy seem ever more real – as if the aim was to make it impossible to tell the difference between the two. However, certain teenagers are in the vanguard in this last respect: they have been trained and then have trained themselves to make their own violent fantasies so real they already prefer to obliterate the difference.

The video games, far more profitable than the movies, are intensively developed to produce an increasingly realistic experience of a first-person point of view *mise en scene*. This appreciably focuses the naturally paranoiac narcissism of youth on the vicious targets out there itching to get him and enables him to kill them first. It's not quite accurate to say that the target is dehumanised by these games – the notion of the human is never even broached. For sure, a certain comfortable level of income, a certain level of us versus them thinking common to all-white suburban enclaves, a certain level of isolation, both physical and mental, a certain level of free unstructured time with few of the available alternatives or stimuli common to urban centres, are all factors that help facilitate the basic training required for schoolboy killers. And then, just about all these boys were actually trained to use guns and lived in homes in which weapons were available and visible. They lived in a gun culture, both in fantasy and in reality, an unbeatable combination.

Taken together, the subjects of this collective profile (vengeful nerds) could be described as psychopathic only if one were to take the radical step of pronouncing what was done to boys in basic training during the Vietnam War as the intentional construction

67

of a generation of psychopaths – 'stimulus discriminators' or not. In any case, Grossman's attempt to historicise the state's refinement of its technology and techniques for making sanctioned killers out of kids directly points to where and when the turn from state training for war to civilian training for violence took place. Unprecedented as it is, the outbreak of schoolyard violence would then have to qualify as one delayed but inevitable spillover or – to use a term popular in the Reagan era – trickle-down effect of the culture's steady abandonment of all its youth – not just relatively privileged suburban white boys. Certain it is that this gleeful abandonment began under Reagan (1980–1988), who successfully preached the gospel of deregulation and government cutbacks and what amounted to welfare for the rich. Known as the Teflon President, Reagan even cutback on school lunches and got away with it. It's only the most drastic and overtly dramatic negative effects of such abandonment that attract the attention of the media. For a few years in the 1980s, gang violence took the lion's share of that attention – without the breast beating about 'America's troubled youth', since young blacks and Latinos were not considered to be representative, just dangerous. The schoolyard murders represent only the outer limit of the abuse effects that our culture routinely visits upon its children every day. The splash in the media, directly related to the splash of blood, measures only that outer limit – preferring to sidestep all these pressurised elements as if they didn't exist. Certainly, the ongoing urban violence against children – much of it gang-related – no longer attracts media attention. We much prefer to look no further than our own white noses – so long as they aren't bleeding.

Much of my own artistic research and production, beginning in the midst of the Reagan reaction, has been concerned with the imagination of violence, the technology of violence, and the consummate effectiveness of training for violence in the American home. In 1985, while living in Southern California, I produced a videotape (with Ernest Larsen) called 'Scenes from the Micro-War'. The videotape depicted a crazed gung-ho survivalist family, played by the members of my own family, who treated their everyday life as if it was an ongoing military campaign. Oddly enough, the idea for the videotape grew

Figure 1
Scenes from
the Micro-War
(1985).

out of our own everyday life. We'd bought a used car that was painted
in camouflage colours – and found while driving around San Diego,
California that the car produced unexpectedly intense responses. The
camouflage coating had transformed our ordinary Dodge Dart into
what was perceived to be a tank. (San Diego, it should be noted, is a
Navy town, with a very large and constant military presence that helps
fuel the local economy.) Following that lead, we chose to reconfigure
every possible visual element in the video in camouflage colours, re-
ordering the visual surface of the video into the landscape of a
domestic battlefield. Early in the video, during a camo-car trip to the
local gas station that's staged like a commando raid, an authorita-
tive voice-over intones the notion that the modern American family
has changed its function – it's no longer merely a consumer unit but
has become a military training unit.

I thought of that deliberately satiric phrasing this past spring
when it suddenly occurred to me that 'Scenes from the Micro-War'
had turned out to be much more prophetic than we could ever have
imagined. Somehow I hadn't quite grasped the inevitable logical extension of the narrative of 'Scenes'. I hadn't quite grasped the extent

✳ to which children are born with few, if any, natural powers to resist the consumption of violence.

In the early years of the Reagan era, with the spear-rattling right-wing in ascendancy, following what was seen as the Carter administration's humiliating capitulation to the 'evil' Ayatollah in Iran, the Vietnam War was reinscribed as a heroic, if tragic, moment for the military. The rhetoric of the Reagan administration, chiming with the American genius for historical amnesia, transformed that vile war into a desperately misunderstood effort to save the 'little yellow people' of a faraway land from the scourge of Communism. With the success of he-man epics like 'First Blood' (1982), starring Sly Stallone as an embittered ex-Green Beret, a vast arsenal of new military technology and artifacts became available in the home, for the first time. A popular poster image of Stallone as the film's Rambo, heavily-muscled and heavily armed, replaced Sly's charming visage with Reagan's: Rambo Reagan. With camouflage becoming a fashion statement, the aggressive anti-war slogan popular during the Vietnam War, 'Bring the War Home' suddenly took on a twisted domestic resonance. For children this meant an enveloping and phenomenally encircling experience in each aspect of their everyday lives: the clothing they wore, the toys, that is to say, the weapons they played with, the necessary consolidation for pre-adolescent and adolescent boys of what constituted appropriate male identities, the uninterrupted and unambiguous parental and cultural reinforcement of that identity. The dawning 'new world order' was eventually underwritten by new family values, a new olive-drab sense of discipline. Major chain stores like Toys R Us, which had removed war toys from their shelves now refilled them with an unimaginable array. All of which is not to succumb to the proposition that war toys inevitably make boys make war. But as another element in the scenario war toys have their place. In Toys R Us, such toys, authentic, i.e. 'realistic' reduced-scale versions of military technology, pretty much took up the whole place reserved for boys' toys. More seriously, what appears to be a trigger is the culture-wide removal of any impediment to the imagination of violence in a significant fraction of American homes.

Inevitably, sooner or later someone would feel compelled to deploy the lessons in killing learned so well at home. In Springfield, Oregon, the 15 year old killer kid, Kipland 'Kip' Kinkel, who first murdered his parents before taking his .22 rifle over to his high school, had been named in his middle school yearbook, 'Most Likely to Start World War Three'. Kip Kinkel made a point of telling his schoolmates that he was ready to rumble but nobody paid him any attention. Then he took it upon himself to act out the unconscious negativity that was always latent – latent not just in him but in the culture itself with the tools that were available in his own home. The Columbine killers, Eric Harris and Dylan Kleebold, produced videotapes in school which dramatised their later attack on their school. Harris had his own web site on America OnLine, which advertised his abilities to imagine potential violence. The day before his rampage, T.J. Sullivan swore he would 'blow up this classroom'. Nevertheless, nobody was able to pick out these explosively violent boys from the pattern of what's long been accepted and expected as everyday violent behaviour common to white male teens in suburban high schools. Which leads one to suspect that these boys, painted by the media as 'desensitised to violence', were in fact more sensitive, more vulnerable to the conditions under which they lived – and unable to handle the burden a second longer.

It is the premise of my video 'Scenes from the Micro-War' that children aren't bad-seeds or natural-born killers, that they have to be trained to become killers and that the home, the haven of domestic space, is the perfect arena to conduct such training, just as bootcamp has for generations been considered the best space to train adolescent males in the activity and technology of state-sanctioned murder. Oddly enough, all too few think of the home as the purpose-built site best-suited to close the electrical circuit between these two kinds of violence: the production of sanctioned violence necessary for the maintenance of the nation-state and the reproduction of unsanctioned violence.

Consider, in this light, the ready availability of murderous technologies in many, if not most, American homes. In Jonesboro, how did 'Mitch and Drew' (as Time magazine referred to them, putting us on a cosy nickname basis with the pint-size Rambos who wore camouflage fatigues during their attack on their school) get the weapons they used on their classmates and teachers? See Figure 2, page 158: Time magazine cover in colour section.

71

Mitch commandeered his stepfather's van which 'contained food, camouflage netting, ammunition, hunting knives, and survival gear.' Then on to Drew's home where, unable to force his father's steel gun vault with a hammer and torch, they 'stole a .38 cal. derringer, a .38 cal. snub-nose and a .357 Magnum that had been left unsecured'. Was that enough? By no means. They drove on to Drew's grandad's, broke in, and swiped four handguns and three rifles, including Drew's favourite, 'deadly accurate' deer rifle. Virtually from the cradle both boys were trained in the use of such weaponry and, as the massacre attests, became expert marksmen, bringing down ten people with 22 rounds in four minutes. So here we have three homes which, as self-contained private environments, functioned both as arsenals and as kiddy bootcamps – and in the end, as time-bombs. The children's training far exceeded the meagre expectations of their parents.

In a situation of war, such disciplined initiative would be cause for commendation, but the children clearly decided that war-like conditions did exist. Children, with their acute consciousness of the seriousness of play, take representation seriously, much more seriously than adults whose abilities to play, and therefore their sensitivities, almost inevitably atrophy over time. Where children sometimes mess up is in their tendency to take play for reality.

And all childlike kidding aside that is precisely the point. These kids lived under the spell of representation, characters in their own self-directed drama. To quote the *New York Times* (May 22, 1998), 'some of the real-life scenes of violence so strongly resemble cinematic events that many experts are convinced that the young killers are not making up the scenarios, but acting out a repertoire learned on screen.' The scenarios they learned on television, on the video cassette recorders, at the malls in the video arcades, while playing their video games, their computer games, while skating on the thin ice of the Internet, are all correlated. They are correlated to a specific use of the imagination, that interior and often treacherous terrain, through which children learn to test the differences and the borderlines between the potential and the real. That specific use could be described as the imagination of violence without consequences, violence with pop-up targets. A schematic version of this scenario is as follows: a young strong super-masculine figure singlehandedly in command of

72

considerable firepower cleverly mows down dozens, hundreds, thousands of his enemies – without evincing the slightest twinge of regret for the havoc he so gleefully wreaks – and without ever suffering more than a symbolic scratch himself. The possibility of empathy is never engaged – let alone dramatised. The sense of power so crucial to the powerless existence of children is reinforced many thousands of times over. But not once is the young imagination asked to feel something for the other, the victim, the sufferer. In fact, the imagination of empathy is deliberately withheld, an essential requirement for any well-trained young killer, whether at home or in combat. The use of powerful (simulated-powerful) tools helps mightily to produce the illusion of power, and is clearly correlated to the consolidation of a particularly pernicious version of masculinity.

If the imagination is a laboratory for psychological experimentation, then these children, adolescents nursing a slight, a grudge, or even a real injustice, as adolescents are also expert at doing, did not feel the delirious and awful freedom of the unhinged aesthetes encountered in modernist literature, they were instead living out a pre-scripted role. For the young adolescent, strung out between the usually authoritarian and inescapably repetitive regimes of the loco nuclear family and the in loco parentis school, it could be a very small step between playing on the surface of a screen and entering it. This small step closes the gap between the virtual and the real – or rather enhances the virtual, eclipsing the real so long as the trigger keeps being pulled, four minutes of forbidden pleasure, of tortured heroism, exchanged for a lifetime of misery – or immediate suicide.

Consider the immediacy of the repetitive activity of playing the video game – in which results are instantaneously measurable and the terrain can with enough obsessiveness be conquered and you can call yourself a winner with no ifs and or buts, a world of boisterous effectivity obviously superior to the taxing pointlessness of school and home. In the video game, the one-on-one claims of the technology become such a relief as opposed to the inescapably social demands of school and home, boring demands based on the deferral of immediate gratification, as opposed to the present-tense of the video game. In the aftermath of Columbine and Conyers, the technology itself often suffered the opprobrium of commentators – given the fact that

Escape Video games offer

Kleebold and Harris were expert players of Doom and Quake, and that Solomon was devoted to Mortal Kombat. But the isolation of a single element in the available technological apparatus is obviously as short-sighted as to single out the killer kids as psycho anomalies. It's the ensemble of relations between the shuttle-like back-and-forth pressures of school and home that the child feels, pressures that he relieves onanistically, repetitively with the joystick, the remote, the push button, the gun. Which is to suggest another way in which the killer kid may differ only in degree not in kind from every other bored depressed, sexually frustrated young adolescent. Which is to say – most adolescents.

The home, domestic space, is the container of these escapist technologies in which the imagination of violence holds all the attention. The home is also a potentially explosive container of psychic pressures which functions both on the imaginary and on the lived levels as a training ground for the ready acceptance and/or propagation of violence on all scales, both imaginary and real, from the not-so-loving pat on the fanny right on up to good old nuclear annihilation. And, while the particular formation in which violence becomes twisted desire that is being evoked here is specific to boys, other formations may apply equally to girls. If some boys kill or maim others, some girls kill or maim themselves, primarily with eating disorders. The grim irony here is that girls are still disempowered when it comes to access to the technology of warfare. The differences between boys and girls lie in the direction of the aggression, in its social acceptability, in its use of the home as a way to keep secrets from the world – despite the avant-garde, rara avis example of Brenda Spencer.

For the past few years, in the U.S. we have lived under the continuous public glare of celebrated criminal cases which revolve around the violence contained within domestic space. Such cases include the virtually endless O.J. Simpson trials; the two trials of the Menendez brothers, young men who brutally murdered their parents in their Beverly Hills mansion; the John Wayne Bobbit cut-up, in which Bobbit's battered wife sliced off his penis and threw it out of the window; the unsolved at-home murder of four-year-old Jon-Benet Ramsey, who had a burgeoning career as a pre-school beauty pageant contestant. Partly as a consequence of these cases, we have begun to admit into our consciousness what we in fact already know: that the home in any class configuration is a deceptive container for explosive violence.

My research into the development of the boobytrap effectively unsettles the stability of every ordinary object in the average middle-class house. U.S. Army manuals are replete with extensive documentation of this transformation of the spatial epitome of safety and security into an arena in which explosive dangers lurk everywhere. The Department of the Army field manual #FM 5-31, published in September 1965, in the midst of the Vietnam War, and simply and arrestingly entitled 'Boobytraps', is filled with depictions of boobytraps and instructions for their construction and deployment. An entire chapter is given over to boobytrapping what looks very much like a suburban home, from the pressure firing device placed under the brick entryway and above the basement windows, to the TV set, the electric iron and even the teakettle concealing sheet explosive, the bed with the concealed pull wire, which will result in a truly orgasmic explosion. (Harris and Kleebold placed 26 such simple but deadly devices around their school.)

For a series of installations on the home as boobytrap, I appropriated these graphic how-to army manuals, and re-constructed domestic spaces in the form of architectural blueprints. Adopting what amounted to a 'Popular Mechanics' do-it-yourself style aesthetic, I then collaged these elements with boy's adventure story graphic illustrations, short circuiting the distinctions among these variously effective fantasy-structures, in the end producing a kind of utopia of destructive possibilities. (See other images Figure 3, page 159, Figure 4, page 160, Figure 5, page 161 and Figure 6, page 161.

A fragile and probably false harmony is achieved through the suspended imminence of complete havoc. What is implied is that at any given moment someone in the home will trip over the invisible wire, sit in the wrong seat, open the wrong window, flip the wrong switch. The disturbing outward serenity of the home continues on the basis of an enormously repressed weight of imminent violence. The style and the tone of the 'Boobytraps' manual and the other army field manuals that I use have spawned at least two popular series of books in the U.S., from the Loompanics Press and the Paladin Press.

Some of the titles from these series that I own are: *Homemade Detonators – How to Make 'Em* (1993), *Backyard Rocketry* (1992), *Death by Deception: Advanced Improvised Booby Traps* (1992), and *Middle Eastern Terrorist Bomb Designs* (1990). Just as the target practice training developed by the FBI has been adapted and marketed as computer games, so the militarily-specialised use of army manuals have migrated into so-called civilian culture in the 1990s. Once again, consumer capitalism heroically bridges the gap between the reality of state-sanctioned violence and the fantasies of home-sanctioned violence, while feeding both on the vastly increased economic power of children and on the supposedly democratising principle of access to information. In fact, schoolboy Kip Kinkel gave a lecture in speech class on how to build a bomb – information probably culled from the same books. Eric Harris downloaded such information from the Internet and then put up his own terrorist-inspired webpage, from which virtual bunker he issued threats against a fellow student. When the threatened boy's parents complained repeatedly to the local police, they were ignored. Who would be so foolish as to heed a teenager's idle threats on the Web? But, after the Columbine massacre, America OnLine immediately and voluntarily supplied the FBI with private information about Harris's online activities, to which it alone still had access. This frictionless alliance between the state and the corporate sector suggests the degree to which the computer itself has been deployed within the home as a tracking and surveillance instrument, supposedly more common to the battlefield. While the Internet has often been touted for its ability to democratise access to information, users of the technology can find themselves highly vulnerable to its anti-democratic exploitation by the state or its corporate clients.

The trauma of the Vietnam War, and that war's technological development (without even considering the no less than 22 other military actions undertaken throughout the world by the U.S. since that conflict) continues to be played out in the domestic arena. Conservative aspects of American home life (involving an assertion of 'family values' celebrating a rigid patriarchal model of the nuclear family that was probably already becoming obsolete in the Fifties) seem also to demand a domestic culture that resembles constant

76

preparedness for war. If the home is indeed a domestic boobytrap – and perhaps we can now agree that it is – then the kind of schoolyard violence we are experiencing now is not an aberration but a logical outcome, perhaps a statistical inevitability. In this model of the domestic boobytrap there is no way to tell when or how the outbreak of violence will occur, only that it will occur sooner or later in some few or many cases – which are intentional effects that heighten the drama and the pervading sense of dread within the space. A sudden fall over the invisible trip wire of rationality is all that is required – and an imagined injury such as those that young adolescents are so adept at sensing would be enough. Or the explosion could be triggered by a lifetime pattern of constant petty or major abuse known only to the inhabitants of the trap, that is, the home. The trigger takes a bit more immediate motivation to cause the explosion, perhaps, but in compensation its consequences can be all the more grotesque and horrific. Sometimes an outbreak can be caused by a distinctly minor event: a child changes the channel on the TV set. An event not even worth noticing. But domestic space is a force field, a power station, a battlefield, and even changing a TV channel could cause a murder, if the circumstances were uniquely propitious. It's important to remember in these charged circumstances that the explosion does not occur by accident. The identity of home and booby trap makes the explosion inevitable. But at the same time the duality pertains: at any given moment the home may in fact just be a home, not a boobytrap. It is precisely this uncertainty that gives the boobytrap its power to intimidate at all times.

Incidentally, David Grossman, who calls himself a 'killologist,' lives in Jonesboro Arkansas, the same small Bible-thumping fundamentalist Christian town as Mitch and Drew, the little Rambos who shot up their schoolyard, and who after they were thrown in jail, asked their jailers to have some pizza sent in, and when they were refused, burst into tears. Theory and practice really are neighbours.

I wish to thank Ernest Larsen for his invaluable assistance in writing this essay.

Shop 'til you[r connection] drop[s]
Considering the electronic supermarket

Angela Medhurst

> *Shops are the most important, single element in most*
> *British town centres. They provide both a vital service*
> *and jobs. Through the payment of rates and the com-*
> *mercial activity which they generate, they sustain the*
> *fabric of town centres and, in addition, provide the driv-*
> *ing force for improvement to the urban fabric.*[1]

Much recent Net analysis and discussion cites electronic commerce
as the technological development likely to have the greatest effect
on UK industry. There are a growing number of e-commerce web
sites that are becoming part of our everyday landscape with the potential to transform everyday activities. Online shopping, in its
relatively short history, has been most successfully utilised by computer hardware, computer software and book retailers. These are
goods that consumers are happy to purchase 'sight unseen'; computer equipment can often be chosen on the basis of a technical
specification and books from a short synopsis. It is only recently
that retailers from sectors where the purchase of goods relies on
personal taste, fit or preference for one item over another because
of size, exact colour or smell have started to develop e-commerce web
sites. My intention here is to examine the experience of the electronic supermarket and to question the impact of online grocery
shopping on the everyday life of women both in the physical and
the electronic store.

Online supermarkets are a reality; if you live, or work, in a trial
area it is now possible to do the 'supermarket run' from any personal
computer with an internet account, then simply sit back and wait
for the shopping to be delivered to your door. At the time of writing
(Spring 1999) Tesco, Sainsbury, Waitrose, and Asda in Britain are operating internet/PC shopping trials to varying degrees around the

country, although none are national and the trials tend to be based in areas of large towns and cities near to major superstores. This relatively recent development provides a context within which to consider how the social and economic geographies of real women are changing as a result of digital services. Much of the recent feminist critique of the digital transformation of everyday life has centred on communication technologies in the workplace,[2] gender identity in on-line environments,[3] the technologies of reproduction[4] and domestic technologies in the home.[5] The online supermarket, however, enables the application of contemporary digital theory to an experience that is both firmly grounded in everyday social and public reality and simultaneously part of a complex economic model based on market share, consumer loyalty and the ever increasing passage of 'Fast Moving Consumer Goods' (FMCGs).

Supermarkets 'occupy a position of peculiar centrality in our culture'[6] and are consequently identified across economic, social, political and cultural theory. In this respect the development of digital services might be important for a number of diverse reasons. The online supermarket has the potential to impact on how – and when – 'traffic' (human and automated) moves through communities, how the 'stores' are supplied and staffed and who within a household does the shopping. Sociological discourse has tended to focus on shopping as a leisure activity, on the purchase of luxury items, electronic goods and clothes for example, while avoiding the more usual experience of everyday shopping for food and provisions. Supermarket shopping has proven problematic when subjected to traditional political and economic analytic tools. In classic Marxist terms household shopping exists outside the recognisable capitalist production/consumption paradigm. Grocery shopping in particular is a task that exists within the broader context of 'housework' and thus forms part of the unpaid labour of the housewife, it therefore serves only to sustain the family – a unit described by Lunt & Livingstone as merely, 'the institution that provide[s] workers for the production system'.[7]

What is evident, however, is that the supermarket as the physical site of grocery shopping has motivated public and political voices; for example, over the effect of out-of-town stores on local communities and in relation to Sunday trading laws. While the campaign to

resist yet another new superstore development continues in my local high street and the existing stores increasingly become '24 hour',
the presence of the blue and white striped 'Tesco Direct' van is increasingly common. The electronic supermarket is beginning to
permeate the physical and social fabric of this area. An apparently
thriving local community served by a wide range of bakers, green-
grocers, butchers, deli's and chemists, but still unable to completely
satisfy the needs of the local residents who receive their grocery
deliveries from the superstore having ordered them online. As Daniel Miller suggests 'we are having to acknowledge that the global
questions of future economic and political development actually
manifest themselves as much in the emotional significance of mun-
dane household responsibilities as in the computer screens of the
world stock markets'.[8]

Reflections & New Realities

An immediate effect of the online supermarket is to bring to the very
forefront of female consciousness the notion of grocery shopping in
the late 20th century. Grocery shopping constitutes a relatively small
part of feminist study outside the broad deconstruction of housework.
Part of the dilemma for feminist theoreticians has inevitably been how
to accommodate the 'useful' aspects of the act of household shopping
into their discourse, while at the same time feeling obliged to consider
it only as a part of the general unpaid labour that comprises house-
wifery. The general significance within our culture, however, of food
and cooking in particular, has undergone a change in recent years. As
the range of available fresh produce has dramatically increased and
eating out has become commonplace, food has become a signifier of
lifestyle and social status. Consequently the supermarket presents a
vast array of opportunities to invest in this new cultural way of life.
Alongside this change in food culture the burgeoning technological
arena has added new status symbols to our lives in the form of mobile
phones and personal computers in a range of sizes from 'palm' to 'desk'
via 'lap'. The combination of chic food culture and digital technology
manifests itself in the electronic supermarket.

Electronic supermarkets present various opportunities to
deconstruct not only the task of grocery shopping itself but also the

potential for more general lifestyle change resulting from new tech-nologies. Initially, for example, it is possible to compare the physical impact of electronic supermarkets on women's lives with the earlier shift in the site of grocery shopping arising from the development of out-of-town superstores. There is no doubt that one of the conse-quences of the superstore has been the degradation of local shopping facilities, once a focal point of the everyday community of women. An ongoing concern of the economists and cultural theorists con-sidering out-of-town superstores has been the impact on the local communities of traders and shoppers. One of the effects of the com-petition created by superstores has been to put pressure on small traders to provide 24 hour services and to provide an increasingly diverse range of goods and services in an attempt to retain local cus-tom. An inevitable result of this has been a change in the people who own and work in local shops. The rise of the 'Asian corner shop' in most large towns can be compared with the transition in staff in the NHS and public transport systems. It has become low paid, hard work that no one else wants.

> The supermarket has not merely replaced a number of lo-cal shops, making shopping for non-durables cheaper and more convenient. The supermarket has grown dramati-cally in the last 20 years, using the technology of the post-industrial revolution, and the impact on local areas through the decline of the small trader has been consider-able. The impact on the domestic sphere is equally important.[9]

Locality shopping facilities serve in one respect to provide a coher-ent everyday space within which to forge social relationships. It is the social relationships that are an intrinsic part of a physical com-munity that have been utilised historically to react against social and political oppression, for example, and consequently to effect change. The gradual fragmentation of local communities (resulting from a long period of Conservative government) and the emergence of the culture of the individual has often been cited as a primary cause of the de-politicisation of the masses.[10] There is no doubt that

81

'shopping for goods remains a social activity built around social exchange as well as simple commodity exchange'.[11] In this respect a shift to on-line shopping has implications for the social relations currently manifested in the physical shopping environments.

One of the most obvious potential outcomes of online supermarkets is the change in human traffic around the physical environment. Shoppers no longer need to travel to large stores via public transport or in family cars, they no longer physically converge in a single space with a common aim – shopping is no longer a shared social experience, either between family members themselves or the pool of strangers sharing the physical space.

How might this change impact on our everyday lives? As Rob Shields reminds us; 'The broader, latent, social function of retail returns us to the historic importance of the market-place as a meeting-place.'[12] So what becomes of this element of 'social exchange' when supermarket shopping becomes electronic and, therefore, remote? Shopping online is essentially a solitary task conducted from the 'private' space of the computer but even when conducted alone the trip to the supermarket is not solitary. The presence of other shoppers contributes to the overall ambience of the store, supermarkets generate a very particular kind of noise resulting from a combination of trolleys clashing, children shrieking and the Tannoy outlining the special purchases of the day. In our contemporary landscape 'consumption has become a communal activity, even a form of solidarity'.[13] In the realm of the supermarket the solidarity often occurs as a reaction to the drudgery of the chore. Shoppers unite in common appreciation of the tiresome aspects of the task: the rising costs, the wilfully disobedient trolley, the checkout queue. Electronic shopping offers no channel for this kind of social interaction at present – although there is no reason why supermarket web sites might not provide, for example, 'chat' spaces as a 'value added' service.

One of the primary objectives for developers of online shopping environments is to build into the service the kinds of benefits shoppers have become used to finding in the supermarket. In addition to a comprehensive range of goods at competitive prices shoppers have come to expect a range of additional services designed to maximise the potential of the space. These include packing services, dry cleaning, film processing labs, crèche facilities, and perhaps most importantly, friendly shop-floor staff available to respond to customer demands and queries. At present the human resource available to online shoppers comprises 'technical support' help lines, available to help with the installation and maintenance of the technology rather than advise on the availability, price or ingredients of goods and services. In this respect online grocery shopping can be a frustrating process. Only when confronted with a list of near identical goods, does the extent to which subconscious decision making has become a necessary part of negotiating the modern supermarket become apparent. Online product catalogues also contain far less information than found on most food packaging. As well as ingredients and country of origin, the symbols and colour codes used to identify, for example, products suitable for vegetarians are missing and descriptions are limited to a few words. In short online cataloguing systems fall well below the level of service that the supermarket customer has come to expect. Although there is no doubt that many of the services provided by supermarkets are merely designed to limit the number of shops we use and increase the amount we spend, the result has been increasingly high customer expectation. 'People want more than products and services; they want convenience as well as a feeling that they belong to something special.'[14]

Online supermarkets also require a different approach to household planning with the participating store web sites often warning of a shortage of prime delivery slots. My experience to date has generally involved a turn around, from placing the order to receiving goods, of four days. Accommodation of this lead-time requires a different kind of planning and forethought than applied to the usual trip to the physical store. It is inevitable that between ordering the shopping and receiving it some items will have to be purchased elsewhere. As well as requiring exactly the kind of trip to the physical shops that the electronic service is supposed to replace this also often means that the goods are purchased twice. This puts additional pressure on those with a limited budget and if the shopper is already paying for a home delivery from the store, numerous small trips to pick up essentials are resented even more. The prospect of simply waiting for the order to be delivered also makes more obvious the ways in which we 'use' the supermarket as more than just a part of a weekly routine of household chores. The grocery shopping is, judging from the evidence of my local store, an activity undertaken by all members of the household. Born out of practical necessity though this may be, it also provides a shared family time, full of predictable, often comfortable tensions and opportunities to engage in family conversations. The electronic store provides no such opportunity and in fact places an additional pressure on the shopper's time planning. The delivery slots are generally narrowed down to two-hour periods, meaning that the rest of the day must be constructed around the wait for the groceries to arrive – and the possibility that they may be slightly early or late.

A relatively rapid outcome of the electronic supermarket might be a shift in the perception of the social status of household shopping. 'One image of the supermarket is that it is the place where basic needs are satisfied through the acquisition of necessities, where basic household expenditure takes place and where women are mainly responsible.'[15] Although still firmly situated in the realm of housework, strenuous efforts have been made by the major supermarkets in recent years to shift the social status of household shopping. Just as the food we cook and eat has become part of a highly developed symbolic system of lifestyle signifiers, supermarkets have striven to reposition their service as a leisure activity rather than a chore. While supermarkets continue to develop environments filled with exotic and desirable goods, the intervention of the personal computer might catalyse the transformation. At a time when personal computers and network technologies represent the zenith of post-modern contemporary life the psychological status of the act of purchasing the goods and services they mediate is automatically elevated. Electronic supermarket shopping thus becomes part of the digital milieu, operating within a different social and economic forum.

Inevitably, as with all discussions around the impact of hypermedia technologies on everyday life the issue of access must be raised. At this stage in the development of the digital network, whose 'everyday life' are we really considering? The possibility of online shopping predicates the availability of the necessary hardware: a personal computer, modem and a suitable internet service provider (ISP). This in itself raises a number of issues. There is no doubt that the required technologies are still primarily within the male domain, whether at work or in the home, suggesting that not only the status of household shopping will shift but also the identity of the shopper. One possibility is that as women have less access to the technology the grocery shopping becomes a male activity. In addition, it is realistic to suppose that the social status of those with internet access is relatively high[16] and as Cope suggests, for online supermarkets, 'the most obvious target audience are the 'cash-rich, time-poor' people who will seek the most convenient time saving route to shop'.[17] At present the online stores are targeting high-income people who work long hours, in fact the Waitrose trials are being conducted entirely through partnerships with corporate employers. Employees at these large, often out-of-town, office complexes can place their grocery order from their desks, with Waitrose delivering the bulk company order to be stored in the canteen kitchen cold stores until the employees collect their order at the end of the day. Outside these schemes it is reasonable to assume that high earners are more likely to have the necessary technology to order online and are also likely to work long hours and be less inclined to spend valuable leisure time visiting a supermarket.

There are however, many people outside this classification who would ultimately benefit from online shopping; the elderly, disabled, rurally isolated and those without cars, for example. The budget UK supermarket chain 'Kwik Save' have already tapped into this market by offering straightforward telephone ordering and home deliveries for visitors to the stores who would otherwise not be able to transport home large quantities of groceries.

As with so many outcomes of digital technologies, the discussion is largely about how we exist within, and relate to, our local environment. Falk and Campbell suggest that the rise of shopping culture and the emphasis on the consumer in economic studies is about 'the restructuring of the relationships between public and private, individual and social, and the "inside" and the "outside"'.[18] Inevitably the applications of new technologies make the restructuring more explicit, encouraging us on many levels to question how we exist within our environment.

Economic Undercurrents

> *Anderson Consulting predicts that by the year 2000, 20 percent of supermarket shopping will be conducted through non-store electronic channels.*[19]

At the time of writing this prediction seems to be a vast overestimate. Although it is very difficult to obtain figures from the supermarkets (the response is simply that the online service is a result of customers' demand) it appears from the few trials that are being run that the refining of the systems necessary to enable significant growth will take some time. It is during this period of refinement and local trials that the form of the electronic store will be decided. The participants in the trials are, probably unwittingly, playing a major part in deciding how the services will develop and who will ultimately benefit from them.

The online supermarket in economic terms can be seen as an example of the move to post-Fordist production models with the supply and distribution systems made possible by electronic retail systems allowing for a greater sensitivity to complex local cultures of demand. Ultimately this may reverse current economics analysis, as Rachel Bowlby suggests, in traditional retail economics, 'the relation between the benefits to manufacturers and entrepreneurs and the benefits to customers is never at issue; it is simply assumed that the two automatically coincide. Critiques of supermarket shopping start from the opposite assumption; that this relation does not hold, or rather that it holds inversely: the more the capitalists benefit, the less the customers do'.[20]

Ideally electronic supermarkets are beneficial to both retailer and consumer, offering the shopper an escape from the drudgery of the task while at the same time enabling retailers to operate from large distribution centres, equipped for rapid stock turn around, without the need for the value added services that store customers have come to expect. In the foreseeable future how realistic is this perfect marriage of supply and demand? As already outlined, access to the necessary technology and the lack of sophistication in the current online systems are prohibitive. In addition, one of the primary drawbacks of most online shopping systems is that they make no economic sense. As Nigel Cope comments, 'the customer is already paying retail prices for the goods picked from the supermarkets. The home shopping companies are then simply adding more costs with the order taking technology and delivery network'. He continues, 'the real benefits will come when the home shopping companies start cutting out the stores altogether'.[21]

It is this dilemma that stands to have most impact on shoppers. In order for the home shopping operators to become profitable it seems likely that some stores will have to close and if the predictions of Anderson Consulting are even mildly accurate this is imminent. Although customers using the electronic channels might welcome the possible drop in the price of goods and a streamlined service, it is those customers without access to the electronic stores who will feel the greatest impact as physical stores close. A possible outcome of this might be a move back to the local facilities neglected because of the out-of-town stores.

In addition, the profile of the supermarket employee is also likely to alter. As Cope predicts,

> *New jobs will be created in areas such as web site design and maintenance, technological support, database management and the answering of email questions sent via the internet. But many of these will be skilled jobs that will be taken by a different set of people from those who are displaced from the stores.*[22]

Supermarkets, with their emphasis on flexible, part time, short contract staff, have a high proportion of female employees. It seems inevitable that without intervention at this stage the skills required for technical maintenance and warehouse work – areas traditionally employing a high proportion of men – will result in the displacement of women in supermarket employment. In my local Tesco superstore shop staff pick the electronic orders from the shelves in the same way as any other customer because the current warehousing system doesn't allow for items to be picked direct. The introduction of electronic shopping has so far had little impact on the employees of the stores themselves, the goods are even scanned and packed at the usual customer checkout in normal store opening times adding to the demand for cashiers during peak-time delivery slots. This system will change if the service is to become economically viable for the stores in the long term however, and the changes will undoubtedly have some impact on the way the stores are staffed.

Form/Function – Some Conclusions

It is still early to be considering the long term impacts of electronic supermarkets. The current trials are limited and are concentrated in metropolitan areas likely to contain relatively high proportions of 'technology ready' consumers. It is possible, however, to perceive the likely route that the supermarkets will take in light of their trials and it is almost certainly true that, 'supermarkets' early online shopping efforts have shown that they wouldn't be able to cope if services suddenly took off with huge numbers of customers'.[23] If online services are ultimately to benefit either the customer or the retailer a restructuring of retail methodologies is required.

The supermarkets appear to be facing something of a dilemma. Having spent millions persuading us that going to the supermarket is a pleasurable experience, encouraging us with in-store bakeries and extravagant displays of fruit and vegetables to buy goods we might otherwise get more cheaply from local stores, the supermarkets are now faced with a technological development that proposes the exact opposite. The 'unique selling point' of the electronic store is that the shopper need not visit the store itself, need not browse the aisles being tempted by beautiful produce but can simply order the groceries, robot like, from a saved list of weekly goods. The currency of the electronic store is radically different from that of the store. It is the technology itself that is currently invested with the status, the kudos, and it is this status that might eventually decide the fate of supermarket shopping. Although there are many elements of the supermarket trip that are unpleasant, frustrating and boring the same can easily be said of online shopping. These electronic frustrations are often nullified by the knowledge that the user is participating in an exclusive experience, part of a small group of people who have

88

access to the cutting edge technology of the moment. As the technology becomes more commonplace the status associated with electronic shopping will change and as with most technological developments the resulting landscape will have changed subtly, not wholesale. Just as television didn't kill radio and mail order catalogues affect the clothes store irrevocably, electronic supermarkets will inevitably become a facet of the broad shopping arena rather than a replacement for what currently exists.

Based on even this brief exposition of key issues it would appear that the implications for the community of women shoppers are many and complex. Online supermarkets engage with a range of economic, social and cultural ideas and consequently have the potential to radically alter the shopping landscape. Without falling into the neo-liberal trap and embarking on revolutionary prophecies it is realistic to suggest that the recognisable geographies that women have learned to negotiate and fought to re-map are likely to undergo a transition in the relatively near future. This transition is already being engineered, appraised and critiqued by economists, marketers and technologists alike. It lies with the existing community of shoppers to ensure that the transition benefits the real everyday lives of women.

The Cyberflâneuse
Metaphors and reality in virtual space

Maren Hartmann

First Careful Steps

> *Walking is the best way to explore and exploit the city; ...*
> *Drifting purposefully is the recommended mode, tramping*
> *asphalted earth in alert reverie, allowing the fiction of an*
> *underlying pattern to reveal itself. ... - but the born-again*
> flâneur *is a stubborn creature, less interested in texture and*
> *fabric, ... , than in noticing* everything. ...
>
> *Graffiti is the only constant on these fantastic journeys; ran-*
> *dom codices, part sign, part language.*[1]

The author Iain Sinclair recently wandered through London as the born-again *flâneur*, picking up graffiti on the way and using this as his interpretive tool to read the city, i.e. the meaning beneath the immediate city space.

At the same time, across the Atlantic, the graduate student Nick Melczarek also walked through 'real' city space, i.e. New Orleans and other U.S. towns, calling himself a *boulevadier/flâneur* and collecting web-site URLs from 'billboards, magazine covers, paper bookmarks, television screens, radio stations and other items (in shop windows, on café tables, posted on walls, wherever)'.[2] His interpretive tools to read the city and its underlying meanings were real-life signs referring to virtual spaces.

I also strolled through spaces lately, picking up signs and using them as interpretive tools. But the spaces and signs I encountered were purely virtual – or so it seemed at first. Franz Hessel, the theorist of the Berlin-flâneur of the 1920s, described the act of 'flâner' as a form of reading of the streets, in which the shop fronts, the faces, the cars etc. become letters and merge into sentences and pages of a constantly newly created book.[3] Like him, I tried to read cyberspace as such a text. But while it seemed to allow me to become a *cyberflâneur*, restrictions seemed to apply to the *cyberflâneuse*. This concept – just like the *flâneuse* of the 19th century – did not seem to exist. The potential similarity between the experience of modernity (the shock of the new in the nineteenth century city), and the experience of cyberspace today (as yet another newly emerging and potentially shocking cultural space) became intriguing. I therefore started another journey – a journey into the relationship between the actual city spaces of the past and their virtual counterparts today. It is a contribution to the imagination of cyberspace.

> *... the Internet can become anything we can imagine and program it to be. It is a most malleable and evolving infrastructure. ... With such plasticity, it is how we think about the Internet that matters.*[4]

The aim is to discover and potentially develop an adequate language to describe the online world(s) and their inhabitants. This will be done in a brief introduction of the flâneur and its cyber-incarnation. I will then move on to introduce the concept of the flâneuse and her potential cyber-counterpart in constant reference to women online.

The Flâneur

> *Main Entry: fla·neur; Pronunciation: flä-'n&r; Function: noun; Etymology: French flâneur; Date: 1854: an idle man-about-town*[5]

> *... the primary traits of the* flâneur, *namely, the detachment from the ordinary social world, the attachment to Paris and the real, if indirect, association to art.*[6]

The flâneur has often been framed as *the* cultural figure of modernity. Amidst all the confusion of the changing world he was the man-about-town with a sense for the true character of the crowd and especially for the meaning of the city. The flâneur is an artist, a dandy,

91

a detective. To flâner is to roam the streets aimlessly and anonymously, mingling with the crowds without actually being part of
them. The flâneur's place was mainly Paris; his heyday was during
the first half of the 19th century.

But the flâneur not only had his 'real-life' existence. He was often
considered to be more of a motif (in literature and philosophy) than
a historical figure and as such had an important influence on our
understanding of the city.

> *The notion of* flânerie *is essentially a literary gloss: it is un-
> easily tied to any sociological reality. It is a marriage of
> several elements including the practices associated with spe-
> cific sites ... and the literary imagination of Paris as a
> metropolis. ...* Flânerie *was therefore always as much mythic
> as it was actual. It has something of the quality of oral tra-
> dition and bizarre urban myth (...).*[7]

The emerging city – mythic and actual – had brought with it the close-
ness of and to other people. They provided a veil between the social
reality of the city and the flâneur. 'This veil is the mass; ... Because of
it, horrors have an enchanting effect upon him.'[8] The flâneur 'sees it
all', but he is also intoxicated by the mass. He is the hero amongst the
masses, but he does not 'see through the social aura which is crystal-
lised in the crowd'.[9] This is partly because he adds memories and other
mythical elements to this experience. He sees an artistically enhanced
true meaning. The fleetingness of the impressions is essential for the
flâneur's perception. Women are included in this as the passing beauty
– necessary for his gaze, but to remain at a distance. His whole being
is about disengagement on the one hand and a certain ability to see
beyond the surface on the other. Both of these attributes are mutually
dependent on each other, but not as easily visible to the outside world.

The Cyberflâneur

> *My name is wjm@mit.edu (though I have many aliases)
> and I am an electronic flâneur. I hang out on the net-
> work.*[10]

William Mitchell, Professor of Architecture at the Massachusetts Instititute of Technology, was one of the first and still is the most prominent self-proclaimed cyberflâneur. This new flâneur-version, otherwise also described as the flâneur in cyberspace, a cyberflâneur, a fast-forward flâneur, a net-flâneur or a virtual flâneur, has been used by some Web-users to describe their online behaviour and experiences. It identifies them as having a certain *attitude towards* and a certain *understanding of* cyberspace: 'The Cyberflaneur strolls through information space, taking in the virtual architecture and remaining anonymous. ... If the Flaneur was a decipherer of urban and visual texts ... , then the Cyberflaneur is a decipherer of Virtual Reality and Hypertexts.'[11]

Most of all, the cyberflâneur conveys the image of someone who is able not only to deal with the shock of the new, but also to use it constructively. He has the air of someone who knows better. He is – again – not part of the crowd, but has an ambivalent relationship to it: he cannot be without it/them, but he will not be one of them. This point is taken up by Stephan Porombka, when he claims that the electronic version of the flâneur is now a cult figure instead of cultural figure.[12] According to him, this figure is seen – and sees himself – as the one who has not been hypnotised by the new media but who is able to reflect despite being a constant stroller of the virtual spaces. The user of the term electronic flâneur as a self-description is creating an exaggerated image of himself, not admitting to the limitations of the online worlds – and himself. This is the veil again: instead of addressing e.g. the inequalities of online access or other problems of the relationship of online to other spaces or his own shortfalls, he suggests that: 'The keyboard is my café. ... the worldwide computer network the electronic agora subverts, displaces, and radically redefines our notions of gathering place, community and urban life.'[13] The cyberflâneur radically romanticises the Internet.

The Flâneuse & Cyberflâneuse

The flâneur's role as cultural (and cult) figure leaves an amazing gap when it comes to the role of a female counterpart. From the 1980s onwards the flâneuse has been receiving greater attention in feminist discourses. Only little or very limited versions of the female variant of this cultural figure were found in the 19th century though, usually related to consumption and/or prostitution.[14] Even bleaker is the picture in the online world (and its surrounding discourses). The cyberflâneuse as such does not exist – at least not yet. Searches do not prove fruitful. Her (re-)creation – based on some ideas drawn from the flâneuse-debate – is therefore the aim.

It is the cyberflâneuse's *attitude* that differentiates her from not only the cyberflâneur, but many other potential online user identities as well (like the surfer, hacker, etc.). Her whole existence is geared towards social reality and contacts with others – this is her particular way of perceiving and dealing with the online world. This is derived – as will be shown in detail below – from the contradictory theorisations of what kind of cultural figure the flâneuse could be. It is partly a mixture of an opposition to the (cyber) flâneur's failings in particular and an identification with women's problems in the online world more generally. But the cyberflâneuse is not only a female category. The point is not an essentialist differentiation between male and female, but a differentiation between flâneur and flâneuse and an opening up of categories that still use some of women's troubled histories in order to create the new.

Women in Public Places

Especially in the first half of the 19th century, women were present in the public spaces, but only particular 'kinds' of women, not necessarily in respected positions. This is reflected by the women in literature: 'Among those most prominent in these texts are: the prostitute, the widow, the old lady, the lesbian, the murder victim, and the passing unknown woman.'[15] This reflects two aspects of one problem: discrimination concerning women's presence was happening in *actual* as well as in *discursive* terms.

The *discursive* discrimination happened on several levels: there was the above mentioned lack of female presence in literature.[16] There was also a lot of talk about prostitution and its negative influence on the city's physical and mental health. Women in public space were blamed for a certain downfall of the city. At the same time 'respectable' women were supposedly saved from increasing dangers by being kept at home.[18] Therefore, the home was created as a new bourgeois haven. Outside, the public sphere – if ever it existed – was created without female participation.

Wilson and Nava partly defy the claim of *actual* non-existence, stating that the spaces women were allowed to visit increased massively at that time. 'One of the most significant changes that took place during this period was a rapid expansion of what counted as respectable, or at least acceptable, public space for unaccompanied women. The category included the great exhibitions, galleries, libraries, restaurants, tearooms, hotels and department stores.'[19]

While this expansion holds true for the latter half of the century, a major point in the discussion is not so much the actual presence, but the acceptable *conduct* in public places. The main attribute of the flâneur (and his cyber-part for that matter) was his disengagement (on the social reality level) and his gaze. And this attitude was still mostly refused to women. The flâneuse is therefore someone who engages actively with the social reality. She turns the refusal into a positive attribute.

Online Presence – Actual and Discursive

In contrast to this, neither access to the – supposedly public – virtual spaces nor a specific conduct were as clearly refused, but for a long time both still posed a problem. Virtual boulevards, at least in Europe, are not gender-neutral yet. While women's use of the new technologies has increased immensely in the U.S.A. in the last years[20] female access in Europe (overall) still seems to lie below the twenty percent mark.[21] These numbers are debatable, but the overall trend has been widely recognised. This is less worrying for the future (by the year 2005 women are predicted to represent 60 per cent of online users – Nua, 1999), but for an analysis of the current state of the emerging culture, it remains important.

The surrounding discourses have reflected the male dominance very clearly. This implies that the Web's initial uses and the surrounding language have been shaped by the male users. Discursive marginalisation showed itself e.g. in the framing of the online world as threatening, because it was supposedly technologically inaccessible for women and full of 'evil' pornography and other harassment.[22] Pornography or sex-related sites are indeed prevalent in the online world and one is likely to come across them at some stage. Commercially, sex is a very important online feature (claiming all the top places in the list of the most used search terms). Sites offer anything from photographs to live-cameras, to information about actual prostitutes and brothels to sex toys. But they also offer prostitutes' networks[23] and emotional support and a great list of other sites that are helping women rather than undermining them. But in the public media discourses these took a long time to be recognised at all.

In terms of harassment, i.e. mainly male verbal violence or abuse, one of the major debates centred on the more extreme case of the rape in cyberspace.[24] Some of women's mechanisms to deal with these complications are gender-disguise or female-only rooms. But like the private spheres of the 19th century, they are separating women from their own presence.

Disguise

The use of disguise – male disguise usually – was one mechanism women developed in the 19th century in order to achieve the same freedom the flâneur enjoyed. Only when she avoided the direct gaze could a woman become an acceptable (because invisible) version of the female flâneur. George Sand, Delphine de Girardin and others were famous examples.[25] The sexual divisions did not allow women to stroll aimlessly and engage in the fleeting encounter. But under disguise, they could be a *flâneur = she*. This creatively circumvented the problem, but didn't solve it.

The same mechanism seems at work again in cyberspace on different levels. Writing about her online experiences, Roseanne Alluquère Stone, who is a woman who used to be a man, describes herself as a flâneur and not a flâneuse: 'I live a good part of my life in cyberspace, surfing the nets, frequently feeling like a fast-forward flâneur.'[26] Referring to the older referent, it suggests the imaginary veil and the exaggerated self-image, which cannot admit to the limitations of the online worlds. She is a *cyberflâneur = she*, which does not incorporate the potential that a *cyberflâneuse* concept would offer. But the cyberflâneur = she is the only trace of female flânerie online so far.

Disguise (in a gender sense) is also a tactic used especially in MUDs (Multi User Domain) and MOOs and IRC (Internet Relay Chat). It is used by both sexes – some people claiming that it helped them to experience the other gender's point of view, others stressing that it serves as a protection from harassment.[27] This provides a creative mechanism, but no solution.

Transgressing Gender Boundaries

The argument that gender disguise partly circumvents the problem does not deny that there are many possibilities for gender-play and non-essentialist notions of gender in the online worlds. But I agree with Barbara Becker in her argument that the descriptive categories in online worlds are very limited due to technological and sociological limitations. Hence the fascination of online identity games is mainly due to the possibility of filling the empty spaces that are left by the lack of complexity, with fantasy.[28] This equally applies to many other parts of the web. There is a similarity to literary texts in this sense – and also to the flâneur's desire for the fleeting encounter which leaves traces only in his imagination.

The cyberflâneuse is trying to bend gender boundaries in a different way, i.e. naming herself in a female category and thereby referring to a whole array of behavioural possibilities that have arisen from the flâneuse's history. As mentioned above, this does not mean that the bearer of the label 'cyberflâneuse' necessarily needs to be female – but he/she/it should want to refer to social reality rather than fantasy, because this is the flâneuse's basis. Games are not her favourite pastime. When Bettina Lehmann claims that there is a specific female way of dealing with the Internet, it underlines this claim (albeit in an essentialist fashion). This female way focuses on social contacts. Women do not consider the new technologies to create distance, but to overcome it.[29] This is the female version of transgressing the sheer usefulness of technologies.

Technology

Another part of the discursive marginalisation today is the claim that women have a strange relationship to technology. They often perceive themselves as having this difficult relationship (despite using many technologies every day), but this is partly because they are perceived as having it. Instead, women seem to have a very straightforward approach to the technology and its uses once they have overcome the initial barrier. But suddenly playfulness, usually considered to be inappropriate for adults, is seen as the right attitude when it comes to technology. Computers are 'toys for the boys'[30] while '... women use the Internet as a tool to help them find quicker ways to do things while men use the Internet increasingly as a toy'.[31] Women can be seen to be altogether more focused on the outcome of online information retrieval than men. Women's search is supposedly more specific, i.e. aimed at finding things rather than pure surfing.[32] As mentioned above, this is different on the social side of things, but it underlines the general idea of the leisurely stroll (= flâneur) still being a luxury accorded to (or rather only asked for by) men rather than women. Women engage and therefore lose the required distance.

'Occupation'

Pornography online has come partly to substitute for the prostitution of the 19th century in terms of the discourse of supposed dangers and an allowed presence of women in public spaces.

> *Prostitution is indeed the female version of flânerie. Yet sexual difference makes visible the privileged position of males within public space. I mean this: the flaneur was simply a name of the man who loitered; but all women who loitered risked being seen as whores, as the term 'streetwalker,' or 'tramp' applied to women makes clear.*[33]

The prostitute's strolling had none of the artistic freedom and the lack of purpose that the flâneur had. He had a purpose, because the arcades not only provided him with a 'home' within an array of life to the outside world,[34] it also provided him with an occupation. He was making use of his aimless strolling as a potential detective or journalist. For women who strolled, this justification did not work. Their presence was 'permitted' as entertainers and caterers, increasingly also as philanthropists. It allowed women to develop a female gaze, looking at other people's lives nearby. 'For Pallier, the female gaze is identification with the subject while the male gaze is the objectification of the subject'.[35] Emphatic engagement – as asked for in philanthropic work – denies total objectification. The veil between the philanthropist (at least if engaged on a hands-on level) and social reality is impossible. Yet again, the female version of flânerie – if it can be called thus – is essentially different from the male.

Art

Another occupation that justified to a certain extent the presence of women in public places was their role as artists. Initially, Impressionist art, created by women, seemed to suggest a remaining restriction of available spaces: 'What spaces are represented in the paintings made by Berthe Morisot and Mary Cassatt? ... : dining-rooms, drawing-rooms, bedrooms, balconies/verandas, private gardens'.[36] But they also painted boating in the park, promenading, and the theatre. Their artistic output was an expression of this opening up of spaces for women and they were therefore framed as being potential flâneuses.

Since artists are more readily seen to be allowed to transgress boundaries, the great number of female artists online does not surprise. The second female cyberflâneur is also an artist: Lucy Kimbell, a London-based artist who works with new technologies. Providing the only female web site about the cyberflâneur, she says: 'The post-modern flâneur has at her disposal the new, non-physical City of Bits, or cities, constituted through the imagination and programming skills of users of the global network of networks of computers known as the Internet'.[37] The femaleness is again only added on: *cyberflâneur = she*. She also adds a very post-modern claim to diverse identities. This leaves the reference to the actual histories behind and takes away their potential liberating function. Kimbell's description also asks for technological knowledge, without which, according to her, the cyberflâneur = she could not survive. Unlike Stone before her, she might not give in to the cyberflâneur's failings (exaggerated self-image and romanticisation of the online worlds), but she gives in to other male notions of the use of technology.

Consumption

The point about engagement and disengagement is best explained through the example of consumption. In terms of boundary transgression, the newly developed department stores of the 19th century were the one area of movement within the public sphere which was approved – even encouraged – for women (especially middle class).

> *The department store was an anonymous yet acceptable public space and it opened up for women a range of new opportunities and pleasures – for independence, fantasy, unsupervised social encounters, even transgression – as well as, at the same time, for rationality, expertise and financial control. … This was a context which legitimised the desire of women to look as well as be looked at – it enabled them to be both subject and object of the gaze, to appropriate at one go the pleasure/power of both the voyeur and the narcissist.*[38]

But like the rational aspects suggest, shopping can also be seen to be an occupation that is neither work nor leisure. It is again not the aimless strolling, or at least not entirely. Moreover, the female does not show the same self-sufficiency that the flâneur embodies, because she shows a willingness to join in the crowd.[39] She is 'unfit', because she actually engages.

> *Women, it is claimed, compromise the detachment that distinguishes the true flâneur. In other words, women shop, and today, as in the early nineteenth century when the arcades first made shopping a new, exciting and specifically urban practice and pleasure, shopping is considered invariably a female pursuit. … No woman is able to attain the aesthetic distance so crucial to the flâneur's superiority. She is unfit for flânerie because she desires the objects spread before her and acts upon that desire.*[40]

The engagement covers objects and subjects – the people, for example, which women encountered through the above mentioned philanthropic work or the less distanced ways of depicting people in their art. Their true meaning of the emerging city and the crowd was a socially real one.

Interestingly, in the online world, businesses seem to have detected a similar attitude in women, because they decided that 'relationships and interactivity are endorsed as being crucial in attracting women to spend online'.[41] It might appear like a strange twist that the most important attributes that a newly created cyberflâneuse would embody are the ones that business picked up as well.

This is a new concern though. Initially, the balance was very different, i.e. men were considered to be the main consumers online. Cyberspace was after all inhabited mainly by men and the main part of online consumption was computer-related material. But throughout the last two years women were discovered as potential online customers, following their role as the main retail consumers in 'real life'. And while men were still considered to be the 'big spenders' online last year – they seemed to spend twice as much (Nua, 1999), partly because they bought more expensive products[42] – a study from December 1998 now claims that women are almost as likely as men to shop online and even spend more money.[43] The crucial distance seems to be slowly creeping back – there might be more cyberflâneurs in the not too distant future.

Fast Forward

The signs that I found on my online journeys, mixed together with a journey back in time, create a difficult mixture. Their use as interpretative tools is not a straightforward one. While there is no cyberflâneuse as such to be found, her potential creation seems to be made up of many additions and subtractions of substantial parts of the original concept.

Taking all these into consideration, the cyberflâneuse seems a pursuing character – someone who does not only lurk, but who acts. She also seems to take a fairly rational viewpoint of her surroundings and herself, not exaggerating anything too much. Being engaged with her surroundings, i.e. missing the flâneur's distance to social reality, she identifies with others online when necessary and does not objectify subjects. She does not take this too far though, i.e. her identification with other people as well as her desire for objects does not lead her to lose her sense of self. She does not identify with the technology as such either, i.e. she does not subjectify objects. She is therefore neither a cyborg nor a postmodern, constantly changing subjectivity. And she does not use the mass as a veil.

This is a far-reaching corruption of the original concept, which could lead to a rather cynical questioning of whether there is anything left of the original flâneur-attributes. What is left is a form of taking possession of space, a perceptive attitude towards the newly emerging, i.e. a heightened awareness. The cyberflâneur and the cyberflâneuse are both engaging and engaged city-strollers who take in a lot of information and find their specific ways of dealing with it. The virtual city is partly a dreamscape and as such it offers itself to the cyberflâneur. As an exciting social space, it offers itself to the cyberflâneuse.

Infocities
From information to conversation

Penny Harvey and Gaby Porter

The Context

In this paper, we focus on a grouping which formed in Manchester, England over a two-year period, 1997–9. These people had been drawn together from various sectors within the city to work on a European project 'Infocities'. The Manchester team formed part of a pan-European consortium to develop pilot applications using telecommunications technologies for local economic benefit. This project was funded under the TEN Telecom (Trans-European Networks) programme of the European Commission's Directorate General XIII.

Infocities grew out of Telecities, a network of European cities. Manchester City Council (MCC), one of the founder members and a core partner of Telecities, put together a successful bid with six other cities for the Infocities project. MCC's advocacy of new technologies stems from a commitment to use them for social and economic regeneration, and to develop Manchester's profile as an international and forward-looking city. Infocities was explicitly aimed at enabling social transformations. The idea was to open access to new information and communications technologies in a framework of social inclusion, and in the hope that the widespread use of these technologies would assist in regenerating the urban economies of Manchester and other participating European cities.

Within the Infocities project, applications were developed in seven areas. Manchester was the lead city for the area of cultural heritage. The Museum of Science and Industry in Manchester (MSIM) was a key subcontractor to MCC in developing a pilot 'demonstrator' to provide distributed access to cultural heritage materials. Other Infocities projects/products included educational materials and networks; transport guides for local people; virtual models of cities.

As the project progressed, the Manchester team became more aware of the extent to which effective use of ICTs was dependent on innovative and flexible working relationships that challenged hard and fast distinctions between providers and users, expert and lay, hard(ware) and soft(ware). Many people felt that there was a potential to create richer communicative experiences both inside and outside the box of the computer, if these kinds of divisions could be broken down. In this paper we aim to show how the technologies not only enable this kind of transformation, but have also been instrumental in a reconfiguration of expertise that has in turn led to changes in the way in which people understand their own working practice. However, not all those in the project understood these changes in the same way and in this respect we have also looked at the gendered and gendering effects of an initiative that was perceived by some participants to be technology led, and by others as primarily user led.

Methods

We both worked in the Infocities project for over a year, on a daily basis.[1] However, to focus on the specific themes of this paper, we carried out a series of interviews with key partners in Manchester. We knew all the interviewees and had worked with them for part or all of the duration of the project.

Guided by seven topics, outlined below, we asked people to think about their own work biographies: how people had become involved with information and communications technologies (ICTs) and how they understood their current contexts of use. We did not ask people to focus explicitly on the Infocities project, as we knew that for most this was only a very minor part of their day-to-day work. The Manchester Infocities team thus provided a focus for a discussion of how people contextualise their experiences of the new technologies, but their answers were not limited to the specific applications developed for the European project. Nor did we ask people to focus explicitly on gender. Nevertheless all the interviewees were aware of our interest in these topics; we asked them rather to discuss these topics as and when it seemed appropriate.

The cast of characters²

Greg (44) the Economic Development Officer for MCC's Economic Initiatives Group, responsible for European networking and certain aspects of policy.

Moira (49) the MCC manager of the Infocities project (one of several projects with a significant ICT element that she manages).

John (45) Technical Manager of G-MING, the Greater Manchester Information Network Group. (G-MING is a broadband metropolitan area network linking the academic institutions of Manchester and Salford. One of the goals of Infocities was to roll out the network to non-academic institutions for public access to the network.)

Tom (50) the project manager for G-MING applications, and the manager of the Infocities ROM project from the G-MING side.

Sparky (38) Director of the Manchester Multimedia Centre, an 'incubator' unit within Manchester Metropolitan University, supporting small and medium enterprises (SMEs) particularly in the cultural sector, and funded largely from European sources. Sparky was responsible within Infocities for overseeing the relationship between the concepts, design and network.

Karen (26) Administrator of the Manchester Multimedia Centre also responsible for providing technical support to the SMEs working at the Centre.

Chloe (26) a designer working at a SME in the Manchester Multimedia Centre. Chloe was responsible for the interface of the Museum online database.

Jane (41) Senior Curator at the Museum of Science and Industry in Manchester.

Ed (24) a web-programmer who was centrally involved in developing the online database for the Museum.

Samantha (45) partner of a marketing company, leading the e-commerce aspects of the Infocities project.

Joy (28) also involved in selling the idea of the Infocities applications, looking for users in the small-business and educational sectors. She is employed by another related project with European funding in Manchester.

Pete (33) an interactive designer working at MSIM. His willingness to develop expertise in computing and networking was essential for the daily maintenance of the new gallery offering public access to G-MING applications and the Internet as part of the Infocities project.

Akis (35) was offering wide-spread technical support, particularly for networking issues, across the Infocities project from a base in the Manchester Multimedia Centre.

The questions we asked were grouped as follows:

- General context of their involvement in ICT (formal study, role models, employment history, motivations, support and opportunities, disincentives, blocks or problems).
- How they characterised their work and the aptitudes required to do it well; how particular skills were acquired; how they evaluated success in their work.
- The relationship between formal training and skills developed on the job; the importance of technical knowledge; the technologies they currently used; unexpected problems or possibilities that these technologies brought up.
- How they characterised their relationship to the new technologies, for example as users, designers, providers and/or promoters.
- How they described their work environment, what relationships they drew on.
- What motivated them: why were they doing what they were doing, why now and for whom?
- What would they change in their working lives if they could change their job or the way they worked?

A Framework for Analysis

These questions encapsulate the ways in which we have chosen to think about 'technology', as skilled practices that encompass relationships and exchanges among machines, designers, producers and users.[3] Technologies contribute to producing people and relations between people, and can thus be thought of as a form of communication.[4] We focused on communicative possibilities rather than the one-way model of information delivery.

Our questions were framed to address what we see as a key issue in current feminist debates on technology: the extent to which the technologies themselves and the skills that they entail are understood to be the preserve of particular and gendered kinds of expertise. There is a perceived absence of women in the design, development and public application of technologies. Various explanations are given for this.

Some have argued that women's participation has simply been erased and made invisible. Female technologists, mathematicians, scientists, have not been acknowledged in the accounts of the development of ICTs, while their male counterparts are hailed as autonomous inventors.[5]

Others have argued that women have been excluded, and offer various ways to understand such exclusion.

i Within a framework where scientific achievements are understood to require social recognition derived from particular political, economic and legal networks, and where engagements with technologies are understood as intrinsically masculine, women find their creative and commercial involvement unrecognised/unrecognisable. Such attitudes also result in women being excluded from the jobs in which they might develop and use such skills.[6]

ii In ICT, a strong emphasis on rationalism and objectivity works to exclude those who do not have the requisite academic training.[7] This point is related to the previous argument that particular kinds of knowledge are gendered. Women develop skills in areas/activities that are seen as less important than male engineering or programming skills. This kind of exclusion was the focus of Cockburn and Ormrod's study of the development of microwave technology.[8] They concluded that:

> to be masculine was to be technologically competent – that is competent either to deploy engineering skills or to manage engineers, to be relatively pro-active, project-oriented, controlling. To be feminine was to have little or nothing to do with engineering, but rather to be dextrous and diligent, to know about people, domesticity and cooking, to be servicing, supportive and relatively available.[9]

Thus, it is argued, given these exclusionary scenarios and the close relationship between technology and masculinity, many women are not interested and/or not willing to engage.

Feminist Responses

Having identified the problem in these terms, feminists debate how they might respond. How might the new technologies be used in the interests of an emancipatory politics? How might women appropriate those areas of social and economic life from which they have been excluded? And is such appropriation desirable?

One obvious response has been to put the record straight and acknowledge the technological expertise of women, through studies of their scientific and technological achievements. Another has been to foster such technological expertise by the removal of obstacles to women's engagement with technology. Centres for women not only offering technical training but also addressing women's fears and disquiet around these technologies have been exemplary in this regard. In their particular ways, both these positions acknowledge the (social) value of technical expertise and seek to challenge the male monopoly on such skills and their social recognition.

An alternative, or parallel, project has been to promote the wider concept of technology and to attempt to reshape ideas of western science and technology, redefining technical skills in a more inclusive way. This project is thus not to recover the works of forgotten women scientists and technologists, nor to train women in what have previously been understood as masculine skills, but rather to redefine those skills which women know themselves to possess as equally valid, and equally technical.[10]

There are tensions in these responses to women's exclusion from technological practice. Some activists and writers have argued that both these positions – acquiring the (masculine) skills and asserting the values of alternative (feminine) practices – should be fought for simultaneously, without worrying about the fuzzy logics that such positions entail. Why not both appropriate the mother-board skills and redefine the keyboard skills that many women already possess? Haraway enjoins women to resist the temptation to choose between positions and argues for the liberatory potential of paradoxical combinations:

> *Taking responsibility for the social relations of science and technology means refusing an anti-science metaphysics,*

a demonology of technology, and so means embracing the skilful task of reconstructing the boundaries of daily life ... It is not just that science and technology are possible means of great human satisfaction, as well as a matrix of complex dominations. Cyborg imagery can suggest a way out of the maze of dualisms in which we have explained our bodies and our tools to ourselves.[11]

Technologies are socially constituted (or shaped), and therefore carry with them the contexts of their design, production and use. In that sense, they may be thought of as gendered artefacts, reflecting social inequalities. However, they also structure (or shape) and produce social inequalities and thus offer strategic opportunities for transformation. If technologies constitute our (gendered) capabilities, this could be for better or for worse. This complex question draws us into the relationships between machines and people in the various contexts of design, production and deployment. We now turn to these contexts through our example of the Infocities project and the experiences and perceptions of those working in it.

The Interviews
All the people interviewed for this study were empowered through their engagement with ICTs. They were all motivated and engaged by developments in this field and were in many respects working with common aims in the context of the EU project. However there were also important differences between them. Age and academic background were obviously important variables. The contexts in which people worked, the positions they held, and the ways in which they used ICTs were extremely varied. We do not therefore make any sweeping conclusions about gendered practice from this small sample of men and women.

The interviews did, however, demonstrate a clearly articulated distinction between male engineering skills and what Cockburn and Ormrod characterise as female 'sustenance' skills,[12] which we will discuss in more detail below. After examining how our various interviewees responded to the questions, we move to some general conclusions in which we consider further this relationship between engineering and sustenance skills in the context of a new technical artefact, the ICT product. Do engineering skills in and of themselves confer more status than "sustenance" skills, or does this new ICT environment allow for interesting reconfigurations of such relationships?

The General Context of Involvement in IT – Motivations and Disincentives

All our interviewees felt that the new technologies had made a clear difference to them. People generally perceived themselves to be working in new ways, in new projects and using the technologies in ways that didn't "fit" into previous categories, largely because a new generation of computers with user-friendly interfaces removed the barriers to non-experts. One issue was how people came to 'reconfigure' their expertise in this changing context. In most cases, they bumped into telematics on the way to something else and had come to absorb them. However, in all cases the technologies clearly represented new ways of adhering to established principles. Thus, with the exception of Tom, engagement with telematics arose despite an initial indifference, or even antipathy, to computing. Many of the interviewees (men and women) had initially thought of computing as 'boring and male'.

Greg was interested in politics: for him, ICT provided a new way to fund initiatives for employment and social inclusion. Joy's career was connected with this through the funding perspective; she was interested in working with cultural industries and drawn in by 'brokering' in the field to create funding opportunities. Sparkey, Chloe, Karen and Samantha were designers who had seen (at different times) opportunities opening up with the advent first of the Macintosh and electronic publishing, then the World Wide Web. Akis had also been introduced to computing in the context of design training, but he'd become interested in writing code to enable him to hack software and database architecture. It was this technical interest in how to change and adapt existing programmes that had led him to work within IT companies, writing software and eventually getting involved in networking technologies.

Neither Moira nor Pete had been trained in computing but both were able to relate to it because of their previous training in science/

maths and electronics. Moira was comfortable with the science and maths behind network technologies, and Pete 'enjoyed tinkering' now that the interface had become more visual and computing was no longer a purely abstract programming activity. Both were able to enjoy knowledge that was a scarce resource in their organisations; for Moira this had the added value of a gendered dimension: 'I enjoyed being in a roomful of men and knowing about IT when they didn't.' Similarly Jane had used computers in the course of her working life, but always as tools for delivering other products and they had never been a central aspect of her work until she began to curate IT exhibits, both in the Museum and online. John had also come across computing in the course of his work, but as a civil engineer this had led him more directly into the computing field than in other cases. He had had to learn to program to support his engineering activities but then had the expertise to move to user support and networking facilities.

Tom and Ed were the only interviewees fully trained in their field. Tom had done an electronics degree and got a technical job in a computing department. However, he thought of himself as a technologist and not an engineer; he had placed himself in what was an emerging field, between computing (the software), engineering (the networking) and users. Ed had read computer science at university.

All the women and Pete stressed the importance of colleagues in motivating their move into new fields. Akis acknowledged the importance of one key figure who had both given him work and taught him a great deal. The other men interviewed presented themselves as autonomous in this respect.

Academic training was generally deemed irrelevant, although those with the most visible institutional positions held university degrees. It was also noticeable that although the interviewees were largely self-taught in telematics, with one exception they had university degrees in subjects which had started them off in the careers through which they came into ICT.

The Characterisation of Work and Skills, and Criteria for Success

The key skills which people expressed were not necessarily in ICT per se. Everyone stressed the importance of self-confidence. This was the first skill mentioned by Greg, and was mentioned by Karen in the context of her effective inter-personal skills. Chloe and Pete talked about not worrying that things might go wrong. Samantha stressed the importance of bravado and of not being put down by men who pretend to know more than you do: "The ones who don't speak are the real techies and they only come out at night!" Ed had a similar attitude to customers and knew he could get away with just staying a few pages ahead in the manual. Greg, Sparky and Tom emphasised their abilities to persuade others to do what they need to do. In Greg's case, this was through public speaking. Both Jane and Karen emphasised that they were not intimidated by the language, they understood enough to know what a particular technology could do, even if they didn't know how. Both John and Sparky felt confident to ask lots of questions if they wanted to find out about ICT developments in their particular fields. Tom also referred to the importance of technical skills in the sense that his job revolved around relating users to applications and to networks, and Akis too depended on technical knowledge and his ability to reconstruct technical processes, fast, in response to what always seemed to be urgent requests for technical support.

Both Ed and Akis stressed that they had learnt their particular technical skills by sitting in front of the screen for hours on end, up to 18 hours a day over several years! They both talked about the tension between technical and social skills that this kind of learning process engenders, and the problems it can cause in both your personal and professional relationships. As Samantha said, "They can talk to something that doesn't talk back."

Sparky described his evangelical belief in ICTs as transforming the cultural industries. Tom and Greg believed that technology-led solutions were ultimately the most satisfactory. Greg compared G-MING to the railway: people used to say it was cheaper and easier to move things around by barge and they could not understand the point of introducing railways. He felt that the infrastructure should be created, and then it would be used. In this sense, his commitment was ideological. John was also committed to the technology-led solution in the more pragmatic sense of needing to know that there is a solution as well as a problem. Sparky argued that user needs should be

112

paramount and the technology should not lead over human beings. He held this belief alongside a very strong commitment to, belief in and love of the technology. This was in contrast to the responses of Moira, Chloe and Samantha, none of whom was enamoured of the technologies in themselves but far more intrigued by their social possibilities. Chloe remarked that the technologies were just tools: they could be liberating, but generally they were exasperating, a view shared by Karen. Samantha found the new technologies exciting but knew that she was fickle, she would move on when something new came up. Moira's attitude was one of 'healthy scepticism'. She said that she should probably just put 'crone' on her door to indicate that she was the older, wiser woman who is interested in people and knows about ICT. She was clear that the appropriate relationship to ICT was about being 'sensible and logical'; in her job she needed to understand both the organisation and the ICT, and not be too enamoured of either. Jane also felt that if you were the kind of person who approached problems in a logical way, then this would facilitate your used of ICTs.

In contrast to the notion of technology-led, autonomous (male) agency, Chloe stressed skills such as flexibility, ability to work to a client's brief and to know who would provide help: "I surround myself with tech-heads." Samantha talked about teamwork, having the energy and foresight to innovate, and taking a client's brief as a challenge. Pete, Ed and Joy in their different fields also used this problem-solving approach to ICT. They started with the problem and sought solutions, using their own imagination and working with others. Akis' approach to technical solutions was distinctive. He felt that his particular skill was in relating particular technological developments to wider economic, social and political contexts, a skill that enabled him to think about finding solutions in relation to the wider picture of the global computer industry. His own experience as a technologist had led him to question the ways in which ICTs are developed (a position shared by Ed who also knew that it is not necessarily the best solutions that survive). Akis thus felt able to understand local problems (that appeared to be technical) in terms of the wider socio-political contexts of the industry.

113

The Technologies that People Use

Most interviewees referred to email, the Internet and word process-ing software as essential tools. Some used databases and Karen was particularly drawn to finding IT solutions in that field. The design-ers were primarily engaged in using and keeping up with the rapid changes in design software. It is clear that basic office skills, such as knowing how to word-process or send emails, do not in themselves create the possibilities for autonomous and effective action. Our in-terviewees did not associate these tools and skills with their expertise perhaps with the exception of Karen who used her wide-ranging knowledge of office software to train others. Generally, there was a sense that these technologies speed processes up, but also that they create backlogs or such a sense of pressure that work may come to feel technologically driven.

When asking people to describe the technologies they used, we were reminded of the complexity of ICT as a web of inter-related and interdependent technologies, with different characteristics and articulations, and differently gendered. The interviewees expressed clear divisions between the office-based technologies (email, databases, word-processors), the design technologies (changing very rapidly, visually engaging, basic toolkit for 'creatives'), and the more mathematically based programming languages and net-working technologies (the domain of the technologists and the techies). Thinking beyond the limits of the Infocities project, the interviewees perceived that the office-based ICTs were in general use, the graphic design technologies were used 50/50 by men and women while the networking technologies and web programmers were primarily men.

How Interviewees Expressed their Relationship to the Technologies

This question differentiated interviewees most clearly. Few were will-ing to identify themselves as technologists and nobody thought of themselves as a techie, although Ed had had a techie phase, and Akis admitted to having been a 'total nerd' in the past. All were clear that the 'techie is a bloke' and a bloke who privileges technical over so-cial skills and doesn't think about the social relevance of what he is doing.

114

All the interviewees saw themselves as bringing technical and social skills together, but in very different ways. We distinguished people's relationships to ICTs in general along two axes – one more functional, the other more emotional, as discussed above. Greg, Tom, John and Joy saw themselves as promoters of (networked) ICT as a process, a way of doing things. They shared a commitment to promoting active usage, but expressed this in more abstract and less purposeful terms than some of the others. John noted that he didn't tend to see 'users' in his job, although the IT environment has changed so dramatically now that 'users' do not require anything but the most basic IT skills, and 'users' are everywhere. Others saw ICTs as tools for building applications. Here again, we found a distinction between ICTs as networks and ICTs as design tools. Akis saw himself as bridging this divide as a "plumber of the future", enabling the systems for the expression of other people's creativity. Akis, as mentioned above, also saw ICTs as tools for mounting a challenge to the ways in which the technologies are developed by large companies. He thus saw himself as an enabler and a provider but not as a promoter of technologies in any straightforward sense. This recognition of the problems within this highly commercially driven sector also worried Jane, and made her uneasy about a strong promoter position.

Several people saw themselves as simultaneously promoters and users. Tom stated the importance of this dual role to keep him in touch with the difficulties of other users. 'There is a danger as a technologist that you think you understand it and it becomes difficult to understand non-technologists' difficulties or reticence.'

Work Environments

Everyone readily acknowledged the importance of working in teams. For Greg, teamwork was conceived in terms of a division of labour. Others (such as Pete and Akis) described the team more in terms of peer support and shared problem-solving – and for Akis this group could be dispersed and accessed online. Tom valued his autonomy. Samantha recognised that, while she herself drew on the peer support of her partner, she was in a position where she provided support for her staff.

Interviewees varied in their views of whether management structures were useful. Some valued the autonomy offered to them in a new field of work. Others felt that they would value more institutional support and strategic direction.

Moira and Pauline raised technical problems in the work environment, specifically the lack of ICT standards within the Infocities project and within their organisations.

Sparky, Greg and Samantha responded to this question by describing the physical environments in which they worked. Sparky regretted that building designers did not consider how people worked with computers, so that the physical work environment was often poor. Samantha had invested a lot of time and money in creating the work environment that she wanted in her company. Joy was concerned about physical isolation from other colleagues because of the location of her office. By contrast, Chloe and Karen worked in a studio environment and both appreciated and enjoyed the immediate support from colleagues that was needed to keep learning within the fast-moving field of web design. Ed felt that the ideal environment for concentration was a small box, but also appreciated the open-plan office space for the contact it provided for those trying to work in teams.

Two of the women interviewees noted the 'gendering' of the environments in which they worked. None of the men commented on this aspect. They either felt that the issue was irrelevant or explicitly stated that it was not an issue in their particular case. Moira mentioned her sense of working as a woman in a male field of expertise.

Chloe disliked working in a 'blokey' atmosphere among the 'tech heads'. She felt that she was treated differently because she was a woman. While she found the other women working with her supportive, she felt that they were criticised by the men for chatting with one another. She was shocked to find that the men didn't talk to one another, even when they met in the pub! Karen, by contrast, working in the same space but in a different capacity, did not think of the environment as particularly gendered. Chloe is a good example of what Cockburn and Ormrod described as the anomalous 'technically competent woman'. Colleagues also identified her in this way, yet noted that she 'hesitates to claim a technological nature for her work'. For

Chloe, the techie is a bloke, and not a particularly pleasant kind of bloke either: his socks smell, he only goes out at night, doesn't talk and is socially incompetent. Given her concept of the techie, it is unsurprising that she chose not to associate herself with them. When she talked about her excitement at a technical breakthrough, or obsession with learning a new programme ('reading manuals in the bath'), she would then laugh at herself and describe herself as 'sad' to behave in this way.

Samantha expressed similar views about techie men, as we have mentioned already. She explained that gender and age are important in ICT design and marketing. She described the kind of woman who is successful: '34/35 is the perfect age because you are old enough to be taken seriously, but can still flirt.' The idea here seems to be that by stressing their femininity women are more able to display technical expertise, for they appear less threatening.

Motivations

Beyond the distinctions between the technology-led and user-led approaches discussed above, there was also a strong sense of the excitement, speed and newness of this field. This is 'untried and untested' technology. There was a sense of enjoyment in the challenge of figuring things out and a sense of play. Chloe even celebrated the fact that the technologies might go wrong. Samantha enjoyed being in a different area, at the cutting edge, of what she sees as a 'nonsense job'. Change also appealed to Greg, John and Sparky: they were interested in the idea of changing the culture of organisations, to adopt new ways of working. Tom loved the technology and therefore loved telling people about it. Moira was more interested in giving people access to information and skills from which they might otherwise be excluded, than in giving them the technology itself. Joy, John and Pete enjoyed the variety of work, the fact that this was a new field with no routine and several people appreciated that they could work quite autonomously. Akis was the only interviewee who felt rather pessimistic about the general direction of change, due to his focus on the ways in which large companies are increasingly able to force people to pay for things that they do not need.

What Would you Change?

This question did not produce answers from which we could generalise. People described issues which were very specific to their current work situation. In general terms, this suggests that, as we mentioned earlier, all these people were empowered by working with the new technologies and derived considerable satisfaction from their working lives. None of them had a burning desire for a change of direction.

Greg wanted more staff; Sparky wanted more managerial support, more funding and less stress. Chloe wanted to work on something less commercial and explore applications for more active users. Joy wanted to move nearer to other colleagues and to manage her own projects. Moira simply wanted more energy. Samantha wished that Manchester could provide the market visibility provided in London (it seems that the Internet is still place-based after all), but acknowledged that there were advantages to working on the fringe. Pete wanted to work more with artists to bring his ICT problem-solving abilities to a more creative and visual field. Tom, the lover of technology, wanted more space and distance from his work; he sought ways to use ICT better, not least to find time to introduce himself to his family. Akis wanted a couple of years to write about the technology and his understanding of how the industry is damaging society more generally. John toyed with the idea of a computer by a pool in Florida but admitted that in the end video-conferencing cannot replace the quality of face-to-face relationships. Ed wouldn't take the computer at all. He would like to make enough money to buy a boat and sail around the world!

General Points from the Interviews

Obviously, there is a difference between knowing what the technology can do, and knowing how it does it. Some interviewees were clear that they did not need or wish to know how the technology worked but were interested in how it could be applied to solve particular problems and/or to create particular outcomes. For these people, engineering and programming skills were not the issue. The skills they mobilised in this process were skills in understanding problems, and particularly in understanding that these are never entirely technical. Even where they appear to be 'technical', problems may be translated into human terms.

Others interviewed tended to work from the frame of technology and to see and articulate issues in terms of technical possibilities and human problems: the fix is technical, the problem is the people who don't understand the technology well enough. In these scenarios, responsibility for problems was placed with the less technically informed (feminine) users, while the technically informed (male) providers were not (socially) responsible: they 'merely' dealt with machines. The problem arises because someone configures something wrong, or unplugs something that should have been plugged in. From this perspective, when a problem arises, the fix does not require mediation between people. For example, when the Museum was having trouble with its network connection to G-MING, the diagnosis was that the Museum's configuration was 'wrong'. The Museum was not invited to take part when G-MING staff were discussing the solution, because this was perceived as technical and therefore not something to which the Museum might contribute. This example illustrates the point that some interviewees thought of themselves as working in an autonomous and self-contained technological domain, while others were quintessentially moving between domains that were, to them, self-evidently human.

If we think about 'power' in terms of autonomy and control ('power over'), the required skills for success in the ICT field are a combination of engineering, software, design and user know-how. However, if 'power' is the ability to get something done ('power of'), then it is not necessary to master all these skills oneself, but simply to know that they go together.

Information Technologies – A New Environment?

It is clear that technological innovations in themselves do not necessarily change gender inequalities, although they may restructure the ways in which men and women work. However, aspects of these new technologies present particular opportunities for a change in gender relations.

We argue that, unlike the case of the microwave oven studied by Cockburn and Ormrod, sustenance and engineering skills cannot easily be held apart with networked technologies. To clarify this point we emphasise, as our interviewees have shown, that ICT practice is extremely varied. We could follow Tom's characterisation of ICTs as bringing together infrastructure, software, communications engineering – and people. ICTs are complex, distributed, interdependent and explicitly relational artefacts.

The dynamic sectors of the industry no longer see engineering skills as the privileged and primary domain of knowledge. As artefacts, ICTs are not tied to the engineering aspects of their production in the same way that other machines and more concentrated systems might be. Once we think beyond the construction of computers themselves to the creation, networking and distribution of applications and sites, engineering know-how does not produce an 'object'. The product of this new engineering knowledge is the link between hardware and software, machines and information systems. Thus, even for those who have trouble thinking in terms of users, ICTs entail multiple ways of knowing. Interestingly, this may make a space for the more 'feminine' skills of managing process.

If we follow the particular breakdown of responsibilities in the Infocities project, those with the engineering skills were predominantly men. They knew not only what the hardware/software/network matrix could do, but how it did it. However, such knowledge in and of itself did not create the product. The key managerial roles in the Infocities project were held by people whose skills lay in using information to link people as well as machines (which includes an important element of technical know-how).

Engineers acknowledge that their skills need supplementing in this way. They need to know about 'users' in order to consider the kinds of applications that might encourage people to use the networks. Of course, artefacts have always required users, but with the complex chain of ICT, demand cannot be identified through undertaking market research to identify who needs a network. The demand for networks can only be created if those networks are carrying information and opportunities that people want, find inviting and may be drawn into. So the people who know how to engineer networks need people who know how to 'build' applications, content and services. This is already a different kind of engineering, requiring input from and interaction with designers and content providers. Applications build on materials and skills that are not the traditional domain of engineering.

Where Gender becomes a Problem

Technological changes, when combined with more general socio-economic contexts, make spaces for more autonomous activities for women. However, the obstacles that women encounter arise from the ways in which gendered inequalities in the workplace have been sustained, despite these technological changes.

- It is disagreeable for women to work with 'techies' – the smelly socks syndrome. Conversely, women do not become 'disembodied' – the physical bodies of men and women get in the way here.
- Institutional hierarchies persist in which one group of people is seen to exercise arbitrary power over others and to prevent successful innovation.

Gender inequality is thus both reconfigured and reproduced through these new technologies, as we have come to expect from previous studies of technology.[13] Acquiring new skills does not guarantee promotion or career progression.[14]

Nevertheless, technological change does create the possibility of collapsing the homogenous blocks of male and female practice and looking instead at the different kinds of masculinity (and femininity) displayed in the contexts of design, production and use. This is suggested by Grint and Woolgar:

> If the deity in the machine is male, if technology in a patriarchal society is essentially masculine, then no amount of reiteration of this point will alter 'reality' – would men really let power slip so easily from their grasp? On the other hand, if the gendered significance of a technology lies in the interpretive framework within which it is constructed, then there is a possibility of deconstructing and subsequently reconstructing the technology.[15]

The difference between technology-led and user-led approaches is important in this respect, as is the related but different distinction between attaching priority to technical fixes or to social relations.

We return finally to Cockburn and Ormrod's advice about reconfiguring women's participation in this new technological context.[16]

i   Women's existing knowledge about technology should be recognised.
ii Women must have an everyday knowledge of how things work.
iii We should all be sceptical of the hype and have respect for other kinds of doing and making.

We end with this point because it re-genders the distinction between the technology-led and the user-led approaches within Infocities. While the technologists might argue that technology-led approaches were ultimately more efficient, they may not have considered whether they were more *effective*. To what extent were the relation(ship)s created by these differing approaches sustainable in the longer term? In the ICT environment, we have become accustomed to information overload, the entrenchment of sexist and racist attitudes and the proliferation of infotainment for commercial ends. We are familiar with the reconfiguration of women's subordinate position in low paid teleworking environments or in the production lines of electronics mass manufacturing and assembly. We must maintain our central concerns and values: who has access to these technologies, what do they provide access to, and how might they benefit users? Ultimately we need to ask how well these technologies met the goals, values and purposes of the society that generated them.

We asked interviewees how they measured success within the project, in order to ascertain their sensitivity to these concerns. However, the Infocities project lacked clear goals and criteria for success, and interviewees' responses were similarly vague and ill-defined: delivering resources and achieving a positive profile for Manchester; bringing together users and networks; 'promoting user needs' or satisfying clients; solving problems. Reviewing the success of the project in Manchester in relation to its goals, the strong focus on technology and relative underinvestment in purposes and values were critical at every level. Manchester's technological vision was that of opening up a powerful network initially created for a closed academic community to a wider community and wide range of uses. This was conceived as a physical task of infrastructure: laying pipes and feeding dark fibre between key sites in the city as the basis for the development of new applications and uses. During the life of the project, it became clear that there was no 'simple' solution of infrastructure. Even within the technological framework, the expectations and requirements of the wider community of Infocities partners for open architectures, systems and services and for assistance and support for users constantly cut across the previous practices, assumptions and priorities of the network. The project was dogged by practical and operational issues such as these, the lack of a common standard for communication and exchange (as Moira pointed out above), and the long delay in delivery of significant elements from a commercial partner in the project. The preoccupation with these concerns limited the impact and spread of Infocities beyond the formal project partners and beyond the individual city. The city subsequently declined an invitation to join another consortium bid led by infrastructure with a range of 'applications' in favour of a strategy of pursuing discrete bids led by users with common interests and needs.

For some, the lack of exchanges and communications with other Infocities partners in the European Union was disappointing. In this important area, nobody took responsibility for engineering and sustaining these relationships, so they did not develop. The trans-European links and relationships were haphazard and thus the technical possibilities were not exploited, or even in some areas explored. Because of the additional obstacles of language and geographical distance, these relationships were visibly weaker and more transient than those within the city itself.

Infocities did, however, focus attention on the importance of values and purposes, and the skills of sustenance. It also increased the knowledge and confidence of participants, especially women, and the potential of other/further fruitful partnerships. Several of our interviewees positively reconfigured the contexts in which these technologies were developed and used in two important ways. First, they did not attribute great importance to the technologies – they were simply useful tools for solving certain problems or creating certain opportunities. The problems and opportunities were interesting and valuable, and the tools were expendable: if these tools didn't do the job, they would discard them and find others that did. These interviewees were committed to the outcomes that could be achieved with the technologies, not to the technologies in and of themselves. It follows that these interviewees were much less concerned about the redundancy of their own skills than interviewees who had a greater investment in the technologies per se.

Secondly, some interviewees sought to reconfigure and redefine the new technologies through creative and art-based projects in which they explored and encouraged communicative and expressive activity with a strong emphasis on social interaction. In this way, the creative possibilities of digital tools, applications and networks may be shared with new users and used to create new dialogues and projects. Such projects did take place as a by-product of Infocities, building on the partnerships created. For example, in the Cyberschools project at the Museum of Science and Industry in Manchester, a multimedia artist worked with five groups of pupils from local schools to make short animations based on the Museum's galleries and collections. Unusually, they were the authors, and offered their own perspectives and stories about the Museum's galleries and collections, which were fresh, amusing and highly original! The project could be shared and extended into their own and other schools using the network connections provided through the Infocities project. G-MING also started to work in close collaboration with the Workers Education Association, Lancashire College, and deaf users in Lancashire to develop an online signing dictionary. This was a pilot project with the aim of making opportunities for the deaf community to explore the communicative potential *for them* of networked information technologies. It is projects such as these which are the most important legacy of the Infocities work, as they are built on the knowledge that human relationships are paramount for the effective use of technical networks.

Windows on the World?
Architecture, identities and new technologies

Jos Boys

This chapter will examine how we conceptualise the relationships between material space and social identity, using the impact of new technologies as a focus. It will propose that some of the ways we think about socio-spatial relationships are problematic, and can obscure rather than explain how 'what we are' interacts with the spaces we occupy. And it will be argued that these ways of seeing are limiting our ability to think through the impact of new technologies on both spatial and social life.

Space is here examined via three, deeply interconnected modes: conceptual space, that is, the shape of ideas we have about things and their relationships; social space, that is, the patterning of our social relationships both mentally and physically; and material space, the socio-spatial environments across which attitudes and actions are played out. I will first outline three problematic 'frames' which shape the spaces of much architectural and cultural theory. These elements will then be explored in turn through discussion of the interplay between a specific place – the South Bank in London[1] – and three new technologies which happen to occur in that location – reinforced concrete, the skate board and the mobile phone.[2]

The three framing elements here explored are, first, the belief that the material world and social identities are in some way an immediate reflection of each other (that is, the social can be directly 'read off' the spatial and vice versa). Second, is the tendency in more recent theoretical work to explore space, technologies and identity by imagining what Haraway calls 'figures'[3] – characters operating as imaginative fictions which disrupt and transgress conventional beliefs and practices through the deliberate 'inhabit(ing) of intersection(s).[4] Third, is the belief that changes in technologies (particularly new communications devices) are having direct and powerful effects on space and identity – literally shifting our sense of ourselves and our relationships to both conceptual and material space.

The underlying argument will be that all these 'windows' on space and identity link the conceptual and the material in a problematic way via representational analogy; that they assume these associative reflections have some explanatory power, and that, in this process authors/designers and audiences/participants' intentions and positionings are obscured. I will suggest instead, that whilst socio-spatial analogies are deeply resonant, culturally embedded 'common-sense' mechanisms[5] for articulating beliefs about both identity and material space, they do not suffice as an explanation for why spaces are like they are, nor how we make and remake our identities spatially. The last section will outline an alternative frame through which to view the relationships between space and identity.

This chapter is based on three single and tiny 'moments' in critical practice and theory, juxtaposed on occasion with personal observations about the South Bank. Its aim is not to offer some unified critique of contemporary cultural and architectural approaches but to pinpoint key problems and begin to ask some difficult questions.

Representing the Real: The South Bank and Reinforced Concrete

The South Bank, the National Theatre and the Queen Elizabeth Hall complex provide an example of a moment in modernism when a new technology (concrete) was used explicitly by avant-garde architects as a mechanism to give physical expression to perceived social attributes in society. Here, I want to unravel the conceptual framework behind that moment. This is not just an issue about one, now unpopular, architectural style but about a structure of ideas relating architecture to identity which, I will suggest, remains firmly embedded in much contemporary thinking.

The invention of reinforced concrete at the end of the 19th century offered the possibility of designing higher buildings with wider spaces between the structural elements, evidenced in new department stores, skyscrapers and office blocks. Like all new technologies, this new element also required a language. Reinforced concrete was a building material like no other – it both had the strength to form a structural, load bearing frame and the potential for an extreme thinness and plasticity of 'skin'. Avant-garde designers explored the expression of these qualities through a variety of means – its whiteness, the ability to differentiate structure from wall, the celebratory expression of wide spans, a playing off of heaviness and lightness, plane and column, rectilinear and curvilinear qualities.

However, although the materials were new, these explorations were framed by a long history within architecture of debates over the relationship between what a building should 'say' and the society in which it was made.[6] By the 1950s in Britain, the belief that architecture should reflect society – and that it must do so 'authentically' – had become commonplace, even whilst there were still disagreements about what constituted an appropriate architectural language, or which specific social qualities should be expressed.

On the south bank of the River Thames in London, various physical manifestations of this modern 'honest' architectural language appear all shaped in slightly different ways by different architects' interpretations of both what constituted an appropriate vocabulary of form and what social values to express. During the sixties, avant-garde architects in Britain became increasingly critical of what they now saw as the *in authenticity* of using white concrete to imply purity (when this wasn't its 'natural' state): against the 'rigidity' of representations of social structure in theoretical works and buildings – particularly by Le Corbusier – and against the 'effeminate' and 'domestic' qualities of Festival of Britain style architecture. These architects looked instead to exploiting the marked surfaces resulting from the shuttering boards in which concrete is cast. This 'in-situ' reinforced concrete is then seen as expressing its own 'raw' and 'authentic' qualities. They were thus deliberately evoking a string of resonant socio-spatial associations, linking together ideas of the masculine, plain, 'poetic' ordinary and 'savagely' noble. These were intended to express British society and culture as a dynamic, democratic and authentic community, through an architectural language of exposed surfaces, concrete layers, walkways and interconnected circulation routes.[7]

The string of cultural buildings alongside the South Bank riverside walk in London (Royal Festival Hall, Queen Elizabeth Hall and Purcell Rooms, Hayward Gallery and National Theatre) were all marked by these processes: most particularly reflecting the shift from the Scandinavian modernism of the Royal Festival Hall to what became known as Brutalism, represented in its later stages by the National Theatre. Here was a deliberate attempt to simultaneously 'read' concrete as a mass produced industrial material and as the new 'nature' for a modern world, physically mirroring the stratas and contours of a roughly carved landscape and beautiful for its transcendental 'realness'. As a history of the building explains:

> *... the reinforced concrete blocks from which the structure would be built would be deliberately roughened with the lines of their wooden casts, to take away the harshness of plain concrete. The concrete, itself, was chosen to whiten, not darken, with age; and in time Lasdun hoped that the building would seem to grow out of the river, slightly stained with lichens.*[8]

The National Theatre and its connected walkway network, then, are classic examples of the belief that the material world and social identities (whether national, cultural, class or gender) are in some way an immediate reflection of each other and that the architect should express specific (appropriate and 'good') social concepts 'authentically' though the manipulation of material form and aesthetics – using contemporary technologies since these most appropriately express the age in which we live. (Figure 1, page 129)

I have shown elsewhere[9] how much of the recent history of architecture can be understood by unravelling the belief that buildings reflect society via the connection between an abstract social concept (community, ordinary heroic, 'rugged grandeur') and it's a metaphorical association in form (flowing, undecorated, 'natural'). I have argued that such an approach is, by its very nature, flawed, since it is structured around an associative relationship which will always be partial, arguable and changing and yet insists on the universality, accuracy and 'truthfulness' of first, 'authentic' association, itself, and second of the specific social concepts and the particular means of their representation selected by different architects in different periods and places.

128

Figure 1 Two women of a 'certain age' going towards the National Theatre; they can't find the way in. They're asking for guidance from another woman, much laughing and pointing. It's the (guidance from) archetypical story of modern architecture in general and the National Theatre in particular.

This, then, is not just an issue of modernism or that moment when the South Bank was being constructed. It is about the structure of ideas relating architecture to identity, the resulting impact on the material world and on the conceptual and material spaces architects and cultural theorists are making in response. At this stage, I will propose the first of three threads in beginning to unravel these underlying conceptual spaces, proposing that in each case there is a whole 'otherspace' which is being rendered difficult to see or articulate.

Breaking the Frame (1)

The assumptions outlined here frame the relationship between architecture and identity through analogy in a self-referential cycle. Two linked concepts – social characteristic and material/aesthetic/spatial form – endlessly appear to reinforce each other whereby the qualities of one are regarded as evidence for the characteristics of the other. Thus concrete (material) comes to stand for all that is 'real' (social). As other writers have noted,[10] such an arrangement makes invisible the conscious act of interpretation by designers and critics in assuming that such a link is valid, in selecting specific social aspects to express and particular architectural elements through which to do it. Both the specificity of the author's position and the potential variations in interpretation by other audiences are thus obscured. Simultaneously, whilst appearing dynamic (via the endless cycling of the loop) this model in fact represents a static space because it cannot incorporate complex transformations or changes. Social and spatial concepts are linked or opposed via the mechanism of a binary relationship, which are either 'good' or 'bad'. Finally, the reflection model relies on representation as *the* mechanism through which the social and the spatial are related, obscuring both its specific qualities as a mechanism of translation and making invisible other mechanisms for articulating our relationships to the world.

An alternative conceptual space to explore, then, is one which sees the reflection model as just *one* of the ways in which we can articulate relationships between the material world and identity. Architectural and cultural theories (and critical practices) need to develop an understanding of how these different conceptual, social and material spaces are constituted, perpetuated and transformed.

Figure 2   Out of the random patterns is a deliberately dynamic ordering of bodies in space.
Individuals walking in pre-ordained sequences backwards and forwards at different speeds, form
a solid box of rectangular air.

To do this, critical analysis of the specific positions of author(s) and audiences has to be a core concern, as do the processes through which all these different spaces are formed and the mechanisms different groups adopt in their attempts to make sense of the world and legitimate their positions in it. (Figure 2)

The National Theatre and its surrounding framework of concrete walkways is about to begin a process of refurbishment,[11] an action in response simultaneously to changing understandings of what constitutes culture (the new centre will incorporate shopping and a broader range of leisure activities), to the perceived continuing lack of popular resonance of its specific architectural expressions of Britishness, modernity or cultural democracy and to the confusing layout and unfortunate undercrofts of a circulation network designed to express the qualities of a dynamic, heroically ordinary and associational community.

Such a concrete act of redesign needs to be analysed, then, not merely in relation to changing architectural forms and their 'reflection' on the social (or vice versa). How could we frame this process so as to explain the changing relationships of the different groups involved across conceptual, social and material spaces?

Feeling the Force? Skate Boarding and Architecture

More recently, and in critical opposition to the perceived limitations of this modernist theory and practice, radical architectural and cultural theorists have been challenging some of the assumptions I have so far described. Many authors and practitioners are examining new ways of articulating social-spatial relationships as an uneven and negotiated process, of making explicit the partiality of the positions of designers and critics and of de-centring representation.[12] I will explore these newer approaches through the work of one theorist who uses the concept of the 'figure' as a means to think about architecture and identity.

In his paper 'body architecture: skate boarding and the creation of super-architectural space', Iain Borden offers the activity of skate boarding as a means to explore 'what happens to the architect and to architecture when critical thinking is rethought as a quotidian procedure, and when appropriations of space, the space of the body, and representations as lived experiences are brought to bear on consciously designed construction as manifestations of philosophy-as-everyday-practice'.[13]

Building on the work of Lefebvre,[14] he argues that skaters occupy architecture so directly through their bodies and actions that they challenge (albeit by limited means) the intellectualisation and over emphasis on visual representation of much architectural space:

> *For skate boarding uses, besides intense vision, a highly developed responsivity to touch, sense, balance, hearing. posture, muscular strength, agility and fluidity by which to perform.*[15]

Borden thus attempts to reinvest architecture with intensity, dynamism and a centrality to the experience of bodies-in-space. This clearly has a contemporary resonance, particularly for critical theory, but it must also be asked why skate boarding is selected to provide the analogy and why are skate boarders the appropriate 'figure'?

Skate boards began life in the United States in the 1960s as a alternative technology for surfers when the waves weren't good enough. They appropriated the man-made and physical shaping of streetscapes as a substitute for the ocean, adapting existing moves and developing new ones in response to this different medium.

Talking to skate boarders it is clear that a preference remains for an appropriation of found space rather than custom made skate parks. This is because it mimics (at some vestigial level) the original surfers' relationship to the sea and because of the endlessly creative engagement between the 'undesigned' conditions of the landscape and the adaptive and transformative demands it makes on the activity. The skate boarders who occupy the undercrofts at the South Bank link immediately and directly into a shared global network with its associated identifying marks and values.[16] Simultaneously they claim and transform a stretch of undulating reinforced concrete and transform it, in Borden's eyes, from an architectural abstraction of 'brutal' honesty into a constantly shifting performative event – as he says, returning the body to architecture. (Figure 3)

Figure 3 The usual scenes outside the National Theatre under the bridge of booksellers and filmgoers at lunch. And then a lone skate boarder, then a few more, passing in and out of the undercrofts. The ground streaked smooth with wear, and marked with use.

133

What role, then, is Borden requiring of this particular figure to reshape architectural theory? Like myself, he too is attempting to operate in the gaps opened up by a refusal of the reflection model. Here, though, whilst not attempting either authenticity or universality, he aims to use metaphorical association to a different end – as a means 'to trouble identifications and certainties'.[17] Like Haraway (see also Sarah Kember, this volume) – although rather more literally – he uses figuration as a means of articulating how knowledge, power and the social (in this case architecture and identity) might be rethought in non-pure and engaged ways. Through the metaphor of skate boarding, the abstract social concepts of the reflection model appear to be made both dynamic and concrete through their embedding into actual (albeit figurative) bodies. For architectural form, the centrality of intensity and dynamism of experience is both a deliberate response to the perceived static and reductivist qualities of some modernist architectural theory and practices and to the inherent solidity and immutability of physical buildings themselves. Somehow, in this process of rethinking is the idea that architecture can overcome the limitations of its structural and material technologies, to also express dynamism and spontaneity. With the emphasis on actual movement, on fluidity and non-fixity, we are now offered not so much the modernist South Bank's 'merely' symbolic and physical representation of dynamic community as concrete strata, high-level decks and walkways, but instead a conceptual analogy between bodies-in-space and an equally unconstrained architecture.

Borden thus maintains a cycle of reflection via representation as the mechanism for articulating both identity and space. What is more, the specific figure he selects reproduces many of the same qualities we have seen valued within Brutalism – even if these are now expressed through a body rather than an architectural technology. Borden's skate boarders equally express radical masculinity, brute poetry, ordinary heroism and dynamic association. Simultaneously there remains a valuing of honesty, authenticity and 'the real' again just relocated to bodies and only by association to the physical forms these bodies occupy. The criticism here is not that the chosen figure exemplifies a certain standardised type of masculine vigour, which can therefore be set against some sort of feminine passivity. It is that

134

concepts such as these – and the mechanism of reflection which is assumed to connect them – perpetuates the conceptual frames of the modernists, whilst appearing to challenge them.

In addition, with the metaphor shifted onto specific human bodies, rather than on abstract social concepts, the social identity historically preferred by artistic avant-gardes (at least in theory if not in practice) seems to resurface. (Values which are usually framed around the bringing together of what would conventionally be kept separate.) As Bourdieu has convincingly shown, the underlying structure of cultural taste generation and consolidation leads artists to tend to prefer either what he calls deliberate transgression or unconventional constraint.[18] The intentional self marginalisation, disruptive slippages and unusual juxapositions of concepts or actions previously kept separate, now offered up by Haraway and others, in fact seems to have considerable similarities with avant-garde conceptual preferences patterned around these two strategies – from the voyeur and beau, through the private dick and the hobo to the cybercowboy and the cyborg.

Borden's writings have a considerable resonance to the contemporary architectural avant-garde, for example, in bringing together notions of the ordinary and the extraordinary. First, architectural space is perceived as being both ordinary (a 'backdrop' to everyday life) and 'found' rather than designed. Secondly, the selection of skate boarding focuses on a particularly intense and immediate relation to the material world. In concentrating on skate boarding, there is a refusal of the possibility that space and architecture might be occupied abstractly. Borden seems to want these 'found' spaces, in their very ordinariness, to be powerful, to be lived not just normally or abstractly but intensely, to be an act of bodily recognition, of celebration and of desire. In the new framework, cultural values may no longer fall into modernists' neat oppositional binaries, but the linking of deliberately contradictory concepts also has a long cultural tradition.

Breaking the Frame (2)

We can begin to see then, albeit in over-simplified form, a sequence of conceptual structures over time. The reductivist modernist model linked a belief in abstract and neutral spatial qualities with the expression of 'progressive' social structures and values in a reflective loop of 'true' (and sometimes determinist) cause and effect. Through the 1950s and 60s, this model went through considerable re-adaptation (particularly by innovative research groups like the Centre for Contemporary Cultural Studies under Stuart Hall in Birmingham) such that both cultural/social and spatial /aesthetic values were now articulated as being partial, contested and deeply affected by the impact of one's 'position' in the process. The loop thus became dynamised as a process of negotiation and transformation over the associative meanings of things.[19]

More recently, with the demise of a Marxist emphasis on 'actual' processes, and a conceptual refusal of ultimate physical 'reality' or coherent theoretical 'truth', the loop linking space and identity through an associative reflection which is also partial, fluid, contested and potentially transformative has more often appeared on the conceptual plane, with considerable degrees of intellectual sophistication and resonance, particularly amongst feminists. In this model (such as Haraway's figures described here but see also, for example, Gillian Rose, Judith Butler and Luce Irigaray[20]) the social 'figure' articulated no longer attempts at any level to link either to real social values or groupings, or to actual aspects of the material world. On both sides of the loop, the conceptual (often called imaginary), material and social 'spaces' we inhabit are inseparable; and the author herself is deeply implicated in, and cannot pretend distance from, the processes being examined. Thus to a partial and fluid associative act is added much richer and 'deep' interpretations of social identity and 'space'.

Yet this model keeps the associative act central to analysis thus containing theory at a level of abstraction which requires only internal and intellectualised evidence. Whilst the author/designer no longer pretends or believes that we can explain the world objectively, rationally and universally, we do not easily give up the power or desire to make sense of the world via those frames inculcated through (and with all the historical baggage of) a particular cultural and creative stance. The preferred values attached by analogy to both bodies and spaces are treated unproblematically (as if obvious) and the linking of values by analogy through binary opposition or association is assumed to be the appropriate framework for relating concepts. Ultimately, then, an uncomfortable question has to be asked of all of us who are involved in architectural and cultural theory. Whose interests are we perpetuating and why?

Here, it has already been proposed that any alternative to the reflection model must ask such awkward questions about the different perspectives and power relationships of different groupings in society and begin to explain how specific patternings of conceptual, social and material spaces come to dominance in specific periods and places and, just as importantly, which spaces are thus obscured or marginalised. Those skate boarders, then, would cease to act as 'figures' eliciting via body-space analogies preferred conceptualisations of architecture and become instead a specific social grouping operating simultaneously across conceptual, material and social terrains, which they occupy alongside, and in engagement with, other groupings (each with their own associated 'spaces' and processes).

Simultaneously, dynamism and partiality need to be located, not in the relationship of concepts but in the conceptual, social and material spaces themselves. Exploring these processes must also accept confusion, mismatch, overlap and contradiction within these 'spaces' and between the different types of spaces. For example, how a conceptual space is articulated through the metaphorical linking of social and material concepts may not align with how social spaces operate in specific material spaces. It is precisely these complexities that an alternative model should seek to make explicit.

The concrete undercrofts at the South Bank were the 'left-over' spaces, the elements that didn't quite fit the architectural analogies between fluid, 'flying' decks and dynamic community. These are now an important location for skate boarders who frequent these as 'accidental found' spaces, paradoxically expressing a dynamic community at its most intense. The undercrofts themselves provide many challenging characteristics which are close to requiring moves designed for the ocean but which need endless creative modifications and complexity. Simultaneously they are a key part of a London-wide route of preferred locations around which skate boarders migrate, itself an endlessly changing and adapted journey and 'found' social and spatial interactions. If this part of the South Bank is a 'no-go' area in terms of resonant architectural representations of social identity, it is nonetheless an integral and important part of both the conceptual, material and social spaces of skate boarding and the processes through which it is undertaken. Interestingly then, we have here a piece of material space which has associative and material resonances, attached to it not through the process of making as its designers fondly hoped – but via a specific grouping's 'spaces'.

But space and identity cannot be explored here without juxtaposing the perceptions and intentions of these skate boarders with the other participants in this particular material space. This is not just the passersby, but also the homeless people who set up a cardboard city in this undercroft many years ago and were forcibly removed, and then, of course the invisible participants in the space, the property holders themselves.

Discussing these spaces only at the level of their associative qualities will fail to explain very much about either how issues of space and identity are being played out in this specific context, or how space and identity might be theorised more generally. The undercrofts are the physical terrain over and through which complex conceptual, material and social processes have been – and are being – played out. These need to be analysed in terms of changing patterns of power and control over the 'meaning' of this location and its various occupants, and over access to the physical place itself. The conceptual, material and social spaces of the undercrofts' different participants do not , however, merely reflect these processes, or provide evidence for explaining them.

At the South Bank struggles over material space have been both secret, and truly fierce. Despite its drab appearance, the spaces have been hard fought over. Homeless people exploited its functional value as shelter, but suffered from being truly demonised figures themselves[21] – in a space which unfortunately supplied popular representational resonances of decay and disorder. This 'visual picture' is in fact unrepresentative of the relatively high levels of social organisation that people here achieved. The conceptual space of that particular analogy obscured rather than revealed the processes it assumed to explain.

Figure 4 Occupying the same space as the tourists and residents, some homeless men and their dogs: long term residents of this park and its benches, part of the scene and simultaneously wary and alert to their difference. An invisible cordon of distancing, an unmarked but clearly defined zone of thin, worn grass between them and the others. Not aggressively drawn from either side, but drawn nonetheless.

Finally though, it was probably the sheer physicality of explicit homelessness (in large numbers) which generated such strong 'lines of tension' between the conceptual, material and social spaces of homeless people and the South Bank promenaders, together with the 'presumed' public, cultural and leisured qualities of the riverside walks material environment. In that context, the homeless were 'matter out of place' and had to go. (Figure 4, Page 139)

Recently, more subtle battles continue with the skate boarders, fought literally over control of the material surface of the undercrofts. There has been a continuous cycle whereby Queen Elizabeth Hall officials smear the concrete with various sticky substances, and then skate boarders reclaim their ground both by finding ways to clean the concrete and by generating new skate boarding moves which deliberately exploit the stickiness. The undercrofts have a less resonant associative quality for this specific group, particularly because here the roof acts metaphorically against an existence perceived as open, 'soaring' and free. They also have both more control and less reliance on this particular place than the homeless did in their time. Still, though, lines of tension flicker and spark, as skate boarders literally and dynamically mark out the boundaries of their action across the paths of the promenaders and cyclists. Their noise, meanwhile, leaks up to intrude on the concert goers above. Here, then, might be a way of describing the patternings of different socio-spatial relations, perspectives and positions of power across conceptual, material and social spaces, where these are not disaggregated from either processes or resource distribution.

Blurring the boundaries? The Mobile Phone

I proposed at the beginning of this chapter that the ways in which we think about the relationships between space and identity have limited rather than expanded debate. I also suggested that we were failing to conceptualise the impact of new communications technologies effectively. As with the other technologies dealt with here, the mobile phone is most often articulated as a metaphor of changes in both society and the spaces we occupy. As Elizabeth Gorsz writes, for example:

> *The implosion of space into time, the transmutation of dis-*
> *tance into speed, the instantaneousness of communication,*
> *the collapsing of the workplace into the home computer sys-*
> *tem, will clearly have major effects on specifically sexual and*
> *racial bodies of the city's inhabitants as well as on the struc-*
> *tures of the city.*[22]

Responses to these shifts tend to fall into the either/or categories I have described elsewhere – a need to control or a pleasure in release.[23] So, for example, the architect and academic Phil Tabor sees a threat to an older, more stable world (in fact a world where representation continues to take place through the luxury of the contemplative gaze rather than via the active and embodied engagement preferred by younger critics).

> *Starting with the phone, electronic media have cracked the*
> *dykes of home and admitted into it all that was tradition-*
> *ally excluded; impurity, worldliness, business, disrespect and*
> *instrumentality.*[24]

Blurring together computer screens and architectural elements, he explicitly offers the analogy of windows on the world as a control-ling mechanism for delineating new boundaries between inside and outside, home and work, public and private – 'whose form neither con-ceals nor arbitrarily represents each new condition, but inherently reflects it'.[25]

Others propose a deliberate immersion in the blurring of bounda-ries generated by new media:

> *It will become increasingly difficult to define the nature of*
> *occupation of physical space; the association between the*
> *space you are in and what you might be doing in it is dis-*
> *solving in a blurring of definitions, and a confusion of the*
> *different activities and roles we assume. This suggests there*
> *may be some tendency for space to lose its defining tags, tak-*
> *ing on a generic nature that is temporally informed as to*
> *its meaning by electronic data.*[26]

141

Whether as a threat to existing and traditional modes of representation or as a means of generating different forms of representation, the new communications media are thus perceived as making physical space in particular, much more fragmented, transient and dynamic. In this model it is assumed that modernist and earlier societies were built around essentialist, mutually exclusive and oppositional categories (man/woman, public/private, home/work) and a simple binary ordering across these categories articulating one as consistently superior to the other. Whereas the post-modern cyberworld is seen as one where these categories have been shuffled and disorganised. In this view, we are now offering a multitude of possible identities which can be treated playfully and where difference between categories loses its hierarchies and becomes pluralist and merely relative. Here, a new set of binaries is generated in opposition to the modernist ones. In its positive mode (where changes are seen as good) particular is set against universal, playful against rational, difference against similarity and pluralism against hierarchy. Where change is seen negatively, the associated concepts are relativist, irrational, fragmented and anarchic respectively.[27]

As outlined before, this framing of the problem also has the underlying structure of the reflection model, based as it is on 'before' and 'after' binary oppositions linking society and space with the mobile phone as symbol either of increasing alienation or a new plurality of expression.[28] Instead, I want first to use the mobile phone as a different type of analogy, that is, one expressing a process rather than a concept; and second I want to suggest that in replacing the reflection model we would be better to conceptualise the mobile phone and other technologies not as analogies at all but as one of the *mechanisms* through which we attempt to articulate coherent and resonant conceptual, social and material spaces for ourselves.

Breaking the Frame (3)
Within the material space of the South Bank complex, mobile phones have the most tenuous and transitory impact of the technologies discussed. Instead, they literally enable us to be in material space and 'not in it' simultaneously.

Figure 5

In Figures 5 and 6 a man and a woman are using mobile phones. But very differently. He is seated, uncomfortably crouching, bending over his phone, sharing a large, abstract sculpture with some children idly playing. She is posed, balanced up on the railing and leaning nonchalantly against a stone plinthed lion, hung against the backdrop of the river. He is focusing on the phone, on the ground surfaces, she on the distance.

In identifying with the woman, I interpret hers as an easy ability to be in two spaces simultaneously – buried in the talk of the phone, and simultaneously aware, performing with her positioning in material space (I've seen men do this too); he is living only in the space of his phone shoulders curved inwards to disengage his body from his immediate material environment. It is as if he is trying 'not to be there – to both not intrude on the physical world, and to be completely immersed in his virtual one'.[29]

Figure 6

Here, what 'matters' about the conceptual, social and material spaces being occupied is their inseparable, dynamic and partial nature. Many authors have dealt with ways of articulating physical space that refuse it as a neutral or transparent medium.[30] Here it is suggested that it is not so much that physical space answers back to social space (and vice versa), as for example, Massey argues[31] but that material space has itself no singularity – it does not exist (or is at least not important) as a thing, only as a process. The definitions of space and objects are thus interdependent at each moment with its location and manner of occupation. As Miller says:

> *The authenticity of artefacts as culture derives, not from their relationship to some historical style or manufacturing process – in other words there is no falsity or truth immanent in them – but rather from their active participation in a process of social self-creation in which they are directly constitutive of our understanding of ourselves and others.*[32]

144

However, just as architecture and other creative actions are about more than the simple expression of identity, so we deal with our need to both survive and to make sense of the world with a huge variety of mechanisms, of which material space is only a part. We thus occupy the physical environment across a range of intensities from concentration to distraction. As Benjamin famously wrote:

> Architecture has always represented the prototype of a work of art the reception of which is consummated by a collectivity in a state of distraction ... buildings are appropriated in a twofold manner; by use and perception – or rather by touch and sight. Such appropriation cannot be understood in terms of the tourist before a famous building. Tactile appropriation is accomplished not so much by attention as by habit. (optical reception) ... occurs much less through rapt attention than by noticing the object in an incidental fashion ...[33]

Here, then, the mobile phone is *not* seen as expressing in object form a dramatic break between old and new ways of occupying space. Instead, to me, the mobile phone offers an interesting analogy for the *conventional* particularities of how we place ourselves in space – that is, operating simultaneously across mental and physical terrains, moving between and across them with different levels and types of engagement and attention. This is not something that has changed from a modern to a post-modern world. We have always done it, whether through daydreaming, story-telling, telephony or the Internet. The mobile phone can thus be seen as analogous for how we might think mental, material and social 'spaces' simultaneously.

If the mobile phone provides an alternative analogy to the reflection model for expressing the ways in which we occupy different 'spaces' simultaneously, it also suggests a means for analysing the inter-relationships of identity and space beyond representation. In this analysis, mobile phone users are not articulated as 'figures' – interpreted as imaginative fictions to disrupt conventional notions of identity and space – but rather as active agents who are engaging with and manipulating a variety of conceptual, material and social *mechanisms*. We are all attempting to 'place' ourselves in the world, in a desire to both make sense of it and to be perceived within in it in particular ways. We can only work with what is available and we do this through a process of adaptation and transformation. Each of the mobile phone users at the South Bank is adapting to a series of contexts, including the material – literally attempting to find an appropriate 'place' in the physical terrain.

Simultaneously, they are 'locating' themselves conceptually and socially in both material and virtual worlds, in relation to their own position (gender, race, culture etc.) and to the others occupying those environments. Interestingly, the woman is deliberately posing – inviting what has been conventionally called the male gaze. Using her mobile seems to enhance her confidence in the physical context, whilst enabling a separation from it. The man, on the other hand (unlike the stereotype of the mobile phone user talking loudly down the line) is attempting to use the physical environment as camouflage and seems uncomfortable, self-effacing.

This, in turn, suggests a way of thinking about the impact of new technologies – particularly communications technologies – on identity and space. I suggest that rather than see these technologies as making identity and spaces more fragmented, multiple and 'on-the-surface', mobile phones are just part of a continually increasing variety of modes available through which identities and spaces (conceptual, material and social) can be articulated. These new additions to our 'toolkits' are not neutral. They come into being through specific processes and in relation to specific spaces. We, in turn, use them creatively – 'building them in' as appropriate to the conceptual, material and social spaces we already occupy. Finally, the technologies themselves have their own specific characteristics through which our attempts to frame identities must be translated.

This chapter leaves many more questions than answers for theory. It proposes that we need a method of analysing positions across conceptual, material and social spaces simultaneously, that we should look to ways of understanding socio-spatial categories without disaggregating them, of delineating the patterns through which power relationships are made and remade, and of exploring the variety of mechanisms for occupying spaces. For critical practice, it raises

146

major, but equally interesting questions about how one incorporates recognition of partiality into one's own design process, how one translates socio-spatial categories into appropriate form-making and how it is possible to engage in decisions over 'meaning-making' which takes places as a contested process in specific contexts. However, until we can shift the frames through which much architectural and cultural theory views the world, we will continue to limit 'how' we can see the relationships between identity and space.

Frontier Dreams I

Stevie Bezencenet

... Then we,
As we beheld her striding there alone,
Knew that there never was a world for her
Except the one she sang and, singing made.

– Wallace Stevens

The Technological Sublime

Is it the recognition of a 'lack'
Should we relinquish our boundaries?
And open ourselves to technological advances?
Bravely enabling ourselves to be everywhere and nowhere?

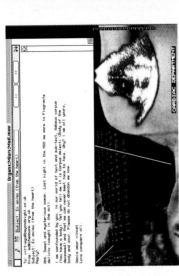

Figure 3

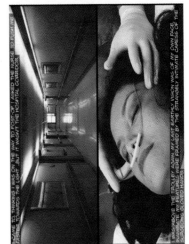

Figure 2

Figure 1

Stills from Jane Prophet's CDROM *The Internal Organs of a Cyborg*. Thanks to PhotoDisc™ for use of stock images.

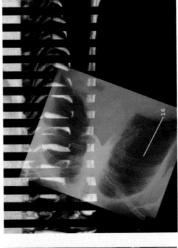

Figure 5

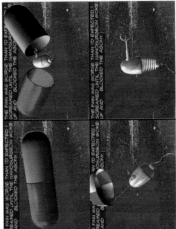

Figure 6

Figure 4

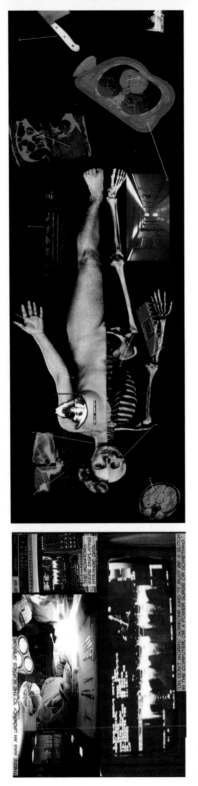

Figure 7

Figure 8

157

Above. Sherry Millner (Fig 2) Cover of Time magazine, April 4, 1998, featuring Andrew Golden as a camo-toddler.

Right. Sherry Millner (Fig 3) Detail: ***"Domestic Boobytrap"*** installation, 1994. Blueprints (with collage elements), and exploded trap-objects.

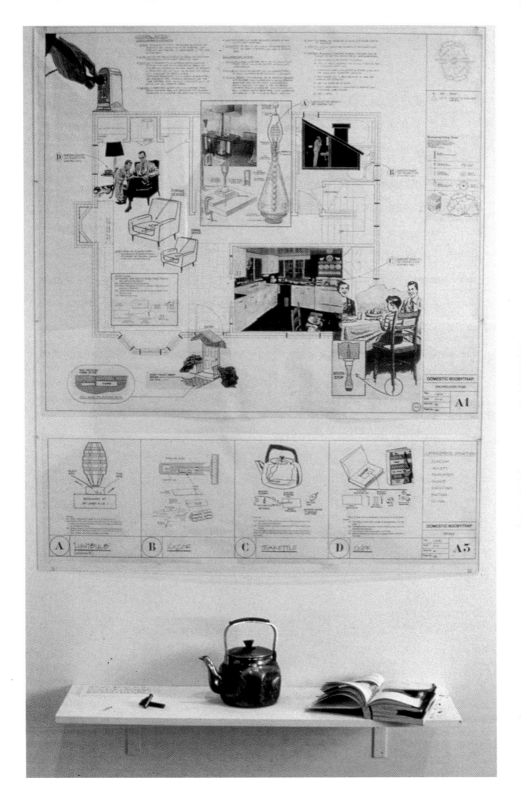

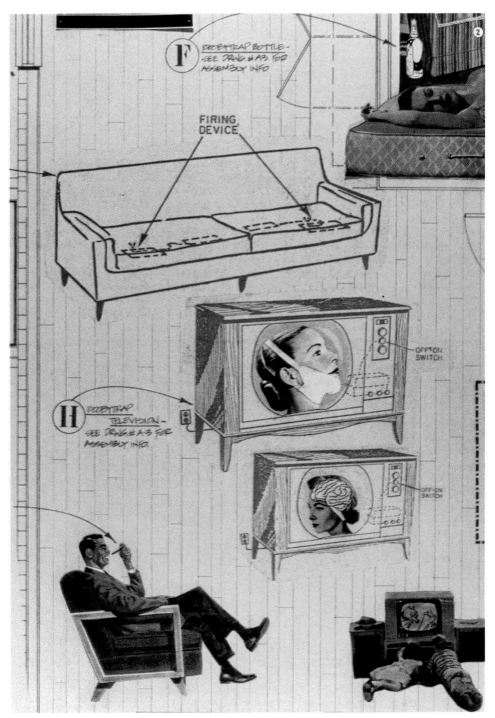

Sherry Millner. (Fig 4) Closer view of Blueprint (with collage elements), *"Domestic Boobytrap"* installation.

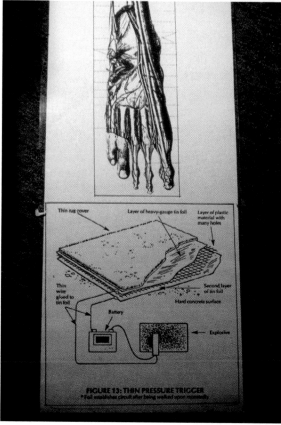

Above. Sherry Millner.
(Fig 5) "This is not a pipe"
Painting. 1994. Detail of
"Domestic Boobytrap"
installation.

Left. Sherry Millner.
(Fig 6) **"Domestic Boobytrap"**
installation. 1995. Blueprint
on floor (beside boobytrapped
staircase construction).

161

162

Gina Birch, *Forty.*

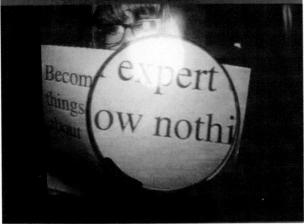

Joan Ashworth, *Eggs, fish and blood (Zero).*

Gail Pearce, *So this is fifty.*

Section 3

Intertextualities
Language, Identity and New Technologies

Women, Computers and A Sense of Self

Erica Matlow

Introduction

> *The things we call 'technologies' are ways of building or-*
> *der in our world. Many technological devices and systems*
> *important in everyday life contain possibilities for many*
> *different ways of ordering human activity. Consciously or*
> *not, deliberately or inadvertently, societies choose structures*
> *for technologies that influence how people are going to*
> *work, communicate, travel, consume and so forth over a*
> *very long time. In the process by which structuring deci-*
> *sions are made, different people are differently situated*
> *and possess unequal degrees of power as well as unequal*
> *levels of awareness.*[1]

In this paper I will be looking at the changes taking place in the working lives of a small number of women lecturers in Art and Design
education, changes that are being brought about through new technologies. Specifically, I will be exploring the relationship that these
women have with their computers, how they have experienced and
negotiated the paradigm shift from mechanical print to the electronic
screen and asking how communication technologies have affected
their sense of themselves.

In writing about computers and women Sherry Turkle argues that
these machines can enable us to think about ourselves in many different ways. Computers can extend the perimeters of our normal
day-to-day lives across a whole range of social relations and cultural
practices. The ways we perceive and experience ourselves in relation
to our working environment, working practices, and relationships
with colleagues, can be changed by computers. Working with computers can also introduce us to new kinds of aesthetics, in art, in

music, in performance and writing, and can augment the way we are able receive the new aesthetics and the potential for being able to interact with them.[2]

Arguably the computer is 'a medium that embodies our ideas and expresses our diversity' and in this sense may be regarded as an extension of the self.[3] This self is never fixed but changes continuously in response to external and internal factors, such as relationships with others, changing circumstances at work and the broader framework of life.

The idea of exploring the relationship women have to their computers, and writing on the subject, came about as a result of a lunchtime discussion with colleagues at the University of Westminster. I asked them about their computers, how they learned to use one and what they now felt about writing with a computer. I then set up an informal process of gathering material based around a series of questions and corresponding by email. My 'sample' comprised twelve women who share certain features in common. Most are engaged in Art and Design teaching in higher education, a few are also artists or work in arts institutions. Each could be regarded as middle-class either by birth or through her professional role. They vary in age from their late twenties to mid fifties. All are English and Caucasian. All the women use a computer at work and/or at home, mainly for word processing tasks such as writing letters, memos, preparing project briefs, writing papers; in addition the computer is used for general research.[4] A few use their computers specifically to develop their artistic practice. None has substantial programming skills nor any significant interest in a specialist aspect of computing. Ten out of the twelve who responded have Macintosh computers. For the sake of simplicity and clarity I have coded the respondents in this paper as R1 – R12.

A small group of respondents such as this can only form the starting point from which to build a case study, however the nature of their answers to my queries suggests there is a rich vein of material to discuss here and to pursue in greater depth at some later point. The twelve women constitute a privileged group, by virtue of their specific social and professional positioning and by virtue of the fact that three quarters of them are known to one another and myself as friends, colleagues, or both. The responses to the specific questions I sent them were often given at some length, they were thoughtful, highly descriptive and discursive. The detail and strength of the material enabled me to position their personal anecdotal accounts, within a broader, general cultural context. In this respect, I had a similar experience to Turkle when she interviewed for her book *The Second Self: Computers and the human spirit.* She too found that when people talk about the relationship they have with their computer they are able to 'talk about things they might not otherwise have discussed. It provided a descriptive language that gave them the means to do so'. In this sense then, 'the computer has become an 'object to-think-with' (in that) it has brought philosophy into everyday life'.[5]

My original plan for this paper was to explore the role of hand-writing and to identify the changes that have occurred for women and their identities in the move from handwriting to word processing. I wanted to know how the twelve women felt about their computer and about their experiences of writing with one. I was curious about whether the shift from handwriting to writing on screen had changed not only the way they worked but their sense of themselves too. I asked about 'sense of self' because I concur with Cynthia Cockburn that ... our 'gendered identity' is constituted not only by these personal histories but also through our social relations' in what Cockburn calls 'the gender pattern of locations'.[6]

I have moved away from the original intention somewhat of considering the shift from hand to computer writing. Instead this paper will focus on the more general key processes that have occurred within communication technologies and explore how these have impacted on women reshaping their sense of self in relation to the individual working practices of each person.

Changing Relationships to Technologies

While the women who responded to my questionnaire have much in common, their experiences, attitudes, ideas and feelings are particular and hence each person gives a distinct account of her sense of self in relation to the computer. Over ten years, roughly the period that the computer has become a familiar part of the work place and household, this relationship has of course undergone numerous changes, which is evident in the range of descriptions of experiences and feelings towards the computer. Some demonstrate a great deal of confidence in their dealings with the new technology, whilst others have adapted with more caution.

According to Dale Spender, most women were probably taught to use a computer by 'one of their female friends',[7] though this was not borne out by my respondents. Most said they were largely self-taught (R4, R5, R10, R3, R12, R7, R2, R6, R8), either from manuals or from watching other people. They also added that their initial experiences with a computer were often ' frustrating and unfriendly' or 'frightening', until that is, they were introduced to a Macintosh computer. There then ensued a different kind of relationship, a 'kinship' with their machine which enabled them to become more confident and develop greater skills. I will return to this 'kinship' later in 'Creativity and Pleasure'. Only R5, and R8, who both use the computer for their art practice, said that they had special training to acquire a knowledge of imaging software such as Photoshop and Macromind Director.

For others, though, working on a computer for the first time opened up the experience of writing as a whole, offering real improvements on previous forms of communication both hand and typewriting, as well as greater ease. R7 found the experience 'challenging and exciting'. R8 found the computer 'alienating at first' but later it became 'familiar and reassuring' and R6 found that it was remarkably easy since it freed her up to 'just get on and do it'. R2, though, was somewhat overawed by the invisibility of the works, by the machine's 'physical opacity', whereas R8 compares her computer to a car, she describes both as tools which should be as transparent as possible.[8] This suggests that she is happy to be able to manipulate the surface simulations of her computer without becoming involved with the machine at any deeper level. Continuing the analogy between car and computer R8 would certainly not want to encourage the fantasy that she could just 'open the hood and see inside'.[9]

For many of the respondents the computer has changed the way they think about how and what they write. Three women, R4, R10, and R12, bought their first computer because they believed that they would not be able to complete their higher degrees without one. R12 noted the convenience on her first use of a computer – 'It was a relief to have a spell checker and to be able to correct 'typos' – unlike the typewriter before which was a pain'. Furthermore she feels the ability to modify text with ease has changed the way she writes, she now writes at greater length and with a stronger critical awareness of the process.

Changing Forms of Communications

For some of the respondents, their relationship to the computer had really only developed when they were already in their thirties and forties. As such, they would be considered by Spender to be the last generation 'reared within a culture in which print is the primary information medium'.[10] These women had grown up and become skilled in a print-based community and had developed certain ways of making sense of their world through print. To this extent they were what print had made them and according to Spender, because of new communication technologies, they now had to change. Certainly, the massive developments that have taken place in communication technology have changed many aspects of the way the respondent group are now able to communicate. It has opened up new ways of working for all of them, created new communities of interest, new kinds of relationships, real and virtual, and provided new roles and new identities.

The telephone is a prime example of this new development. The telephone has been available as one of our prime means of communication for the last ninety years but with the development of email and the Internet, our use of the telephone has multiplied. 'Nattering' on the net has become as much a part of our lives as perhaps the telephone was to our mothers.[11] Spender notes that talking on the telephone is considered to be one of the most common 'outside of work' activities for women. The act of telephoning, through a modem with email, can also extend a woman's working life as it enables them to build networks in cyberspace, and to converse through bulletin boards and mail lists. It also enables them to pursue their research on a global basis with the use of search engines.

Research has shown though that more women need to be enabled to use the net and to become directly involved in it so that they can then shape it to their own needs.[12] Another disadvantage is that although the telephone has extended the means by which women can communicate, it has also extended their work into their leisure time.

Writing is another important form of communication which has also been radically transformed by the advent of computer technology. With the application of word processing software, it is now possible to write and 'announce' ourselves using the different registers of formal or informal presentation and language. Text can now be manipulated in many different ways, instantly on screen. Changes can be made to the parameters of the internal writing space of the screen and the text font and size enlarged or reduced to the extent that our relationship with the design of text is 'now a malleable and self-conscious one'.[13] In this way the computer has rapidly changed our experience of writing. The computer interface presents us with an interface which substitutes 'iconic representations of reality for the real' making it possible to put our thoughts into writing actions seemingly as part of a natural process.[14]

If the computer has changed the way we write has it also changed the way we instigate the writing process? The respondents were divided about this. Some chose to draft out their initial material by hand; R3 R5, R7 and R12 started on paper before going to the computer to translate their thoughts to screen. R4 though, has always used the computer, unless she is writing something very personal, and, like Turkle, finds the computer utterly compelling.

> As I write these words, I keep shuffling the text on my computer screen. Once I would have literally have had to cut and paste. Now I call it cut and paste. Once I would have thought of it as editing. Now with the computer software, moving sentences and paragraphs about is just part of writing. This is the reason I now remain longer at my computer than I used to at my paper writing tablet or typewriter. When I want to write and don't have a computer around, I tend to wait until I do. In fact, I feel that I must wait until I do.[15]

Using the computer has enabled R8 to write much faster and clearer, since her initial ideas are often jumbled and half formed, moving chunks of text around means she can organise both thoughts and writing on screen. This proves a creative way of writing as she can hone an idea which starts off semi-formed. Previously she wrote on bits of paper and literally moved them around on the floor! R9 also writes what she describes as a form of 'automatic jumble' which is then sorted out on screen.

Using a computer to write with has also meant that our handwriting is being affected to a greater or lesser extent, by subtle changes of purpose and perhaps status. Nearly all of the respondents commented that their handwriting had deteriorated considerably since starting to use a computer. R3's handwriting was now 'almost totally illegible' and R12's was 'less well formed'. R9 stressed that she considered that the 'physical act of handwriting as (still) very important', whilst R8 thought it was something 'pleasurable and special'. For most respondents, the activity of handwriting was now being reserved for the more intimate and expressive forms of communication, letters to close friends, postcards or writing poems, scribbling aide memoires or jotting down ideas. The acts of handwriting in these instances were not for public consumption but had been moved to the more private sphere of these women's lives.

Writing on a computer screen has enabled us to visualise text in three and four dimensions, moving around text as if it existed in a physical space imbuing the words we use with a particular sense of location and creating information environments that enable us to negotiate with textual information from within a range of different dimensional, virtual perspectives. A striking example of this is the work of the late Muriel Cooper of the Visible Language Workshop at MIT.[16] Her work, for which she coined the term 'Information Landscapes', recreated a sort of dreamscape 'where clusters of information objects are scattered throughout the three-dimensional space on a computer. The viewer can fly through this space and encounter the separate intelligent objects and not get lost. The environment provides context. For the moment, these objects are discrete groupings of data – small systems that tell us about air traffic control problems, mutual funds, or automatic text layout, for example – each placed in a new dynamic environment'.[17]

R1 said that being able to visualise information in a three and four dimensional environment, like the 'Information Landscapes' above, meant that she could now make connections between physical and conceptual spaces on screen and in her head, in ways that previously had not been possible on the two-dimensional surface of print based media. To quote from Lanham, 'All kinds of conceptual relationships can now be electronically modelled in dynamic and compelling forms'.[18] The virtual space of the electronic screen has enabled her to break from the familiar linear sequencing of text to finding original ways of putting text together with different voices, emphases, hierarchies and linking structures using the language of HTML and the Internet. We have moved from the linear regularity and standardised format of print to the moving 'unpredictable, multiple and non-sequential' aspects of the electronic medium.[19] Though this medium can deliver 'compelling perceptual experiences'[20] and can extend the tradition of television, film, photography and even representational painting, some consider books still have the advantage with respect to computers. 'Even if printed on acid paper, which lasts only seventy years or so, they (books) are more durable than magnetic supports. Moreover, they do not suffer from power shortages and blackouts and are more resistant to shocks.'[21]

Work and Identity

The respondents in this chapter all have a strong sense of self in relation to their work at the University, either as full time lecturers, visiting lecturers or researchers. Their work identity is constructed around many different kinds of discourses, conceptual, material practice, relations with their staff and students and with management. In the course of their work they are involved with many activities, organising new course options, giving lectures, delivering new projects, curating an exhibition, organising a conference or whatever. Most of these activities require a high level of intellectual expertise, organisational ability and management skill and all respondents talked about how the computer had assisted them with this work. R12 relied on her computer to help her at work, it had enabled her to become more efficient and because she worked primarily with text she felt strongly she could now not do her job without one. The computer also offered a public face to the outside world, a formal connection to a global professional academic network.

Most respondents had either a computer in their office or had the use of one which performed a different role from their computer at home. The computer, and in most cases this was a Macintosh, enabled women to feel both efficient and businesslike. This attitude must be viewed in direct opposition to the marketing of Apple computers. which has, strenuously, from its beginning, emphasised the 'image of being a counter-culture' to the world of work.[22]

It is evident from the responses received that the computer has made it possible for women to feel much more in control of their written work. The move from the printed page to the computer screen has transformed the act of writing . It is not just what or how it is written but also the look and the overall presentation of the words on the page that have changed. It is possible now to achieve a (professional) standard of output and presentation at low costs previously only possible with mechanical print. Now it is easy to get high quality work, and the laser or ink jet printers that sit on our desktops can produce printed material that looks smart, neat, and efficient. In this respect, both R7 and R9 find their computers very useful tools enabling them to produce formal things like letters to solicitors or grant applications, though R2 believes that computers haven't really altered the writing of 'decent correspondence', as the activity of writing, regardless of the tool one uses to write with, has always been linked to a particular function and purpose.

Words can now be manipulated and moved around to create different relationships that transform our understanding. Texts can be produced, with the application of different software packages and programming skills, that are now highly individualised and appropriate to any particular circumstance. Typing skills are no longer the prerequisite to producing professional looking documents when even the basic concepts of typographic layout are embedded in the word processing software as part of its default system.

Two respondents, R2 and R1, were critical about the way a computer tied them to a desk, they both preferred the informality of a laptop as it offered freedom of movement and greater flexibilty in terms of how they worked. It is interesting to note that the degree of informality expressed by these women in the relationship that they have, or would like to have, with their computers, is in stark contrast to its original designed intention, as a calculation machine.

R1 sees her computer as a definite 'substitute for typing and not for handwriting or sketching' and that electronic writing offers real improvements over writing for print. Like Turkle, she believes that the computer has offered us a flexible editing facility, cutting and pasting, correcting and moving type around on screen, which she considers to have been 'an amazing bonus'. The computer has also 'enlarged our thinking spaces' and, with its 'dynamic layered displays', offered up 'endless possibilities and choices'.[23] Though as the Internet increasingly becomes our prime platform for dissemination and communication so our control over how we present ourselves on screen has decreased. As R1 states, she has lots more choice and control over what something looks like on screen than she did when working in print. Paradoxically, once her work leaves the confines of her screen and is published as a web page, she relinquishes control because users can set their own preferences and styles which override the ones originally specified in her design.

For most of the respondents the computer has given them a greater flexibility in how they approach and conduct their work in relation to preparing texts, and the introduction of the computer into their offices is considered a positive move by most respondents. On the other hand, because lecturers now have to handle their own word processing there is the assumption that they no longer require help from the secretarial staff, which then puts an extra work load on the lecturers, an aspect I will take up in more detail in the later half of the next section.

Gender and Class Shifts: Reframing Authority

Over a relatively short period, approximately ten years, the context and activity of the work of the University, such as teaching, research and administration, has changed significantly from an emphasis on the printed page and book to a focus on the computer. This has brought with it significant changes both to staff members' identity as lecturers and to some degree, their authority. Traditionally lecturers have been the conduit through which learning has been filtered with books as the prime depository of all knowledge. Now lecturers no longer rely exclusively on the printed word for the knowledge they need to teach with, and can direct students to particular points of reference with considerable precision by citing a particular web site address. The World Wide Web can make available the entire contents of hundreds of different libraries around the globe that can be accessed, instantly. The authority of the printed word, at one time the exclusive priority of the intelligentsia, can now be participated in and accessed by a mass audience and mass readership. How long will students go on needing their teachers when they can use a computer to access more appropriate forms of knowledge, instantly? When students have the possibility of being networked across all departments in their own university to all other universities, nationally and globally, to enable them to discuss that knowledge with other staff as well as with their peers?

How have these developments in the communications technologies changed the authority of University lecturers? My respondents have certainly had to learn different skills and new ways of working in relation to the computer and in doing so have defined new roles and ways of working for themselves. Dale Spender talks about how the 'certain values that have served us well while we used print to make our world' can now actually be 'an obstacle – now that we are obliged to work with the new technologies'. We are continually having to learn new skills, it doesn't stop with word-processing, we now have email and the net and 'for those of us who were reared with print, the continual effort to learn the new technologies will be an ongoing fact of life.'[24]

An example of the change in a lecturer's authority and the subsequent shift in the teacher/student relationship is indicated by this example from R7 when she admits that because she has little actual knowledge of how a computer works, she often elicits the help of her students. 'The advent of the computer has upset traditional images of pedagogic authority ... if education was once built on an apprentice system that included a careful mastery of tools, that system has been severely destabilised. Many students are now far more advanced in their control of their electronic tools than their instructors are.'[25] It is not only the students that set the new standards for technological achievement, as R3 acknowledges when she says that she is neither as confident nor as 'effortlessly experimental' as her five year old son in her relationship with her computer. As Spender explains, being computer-competent is not all that difficult, 'three-year olds can manage the new technologies, often with much more facility than can their parents'.[26] She goes on to explain that this is not necessarily because they are more technically gifted but because they have grown up with it, for them technology is 'just the way the world works'. For the rest of us though, and also for most of my respondents, 'technology is what wasn't invented when we came into the world'.[27]

An example of another role change is expressed by R5, who, as an older woman, proclaims that she can hold her own in what is often a young man's world and feels proud of it. Here she counters Spender's assertion that 'The world of computers and their connections is increasingly the world of men: as more research is done in this new area and more findings are presented, the more damning is the evidence. Men have more computers, spend more time with them, and are the dominating presence in cyberspace. Considering its roots are sunk deep in academia and the military-industrial complex, that's hardly surprising'.[28] R8 also demonstrates an ability to challenge what is perceived as the predominantly male environment of the computer, by sorting out her own software problems, despite her earlier claim stating her desire for the transparent computer, here she has been able to dive beneath its surface and to show self sufficiency.

The introduction of the computer for word processing transformed what had hitherto been considered 'women's work', typing, and with it, a change in gender relations and people's attitudes.

Word processing though 'cannot be treated as just another piece of new technology, it affects a specifically female area of work, and for this reason the impact of any job loss or reorganisation is likely to be subtly, though significantly, different from that in traditional male areas of work'.[29] Certainly we have witnessed a considerable shift in the kind of work that lecturers have had to incorporate into their busy work schedules. The age of the hand written project brief has (almost) disappeared to be replaced by the word processed document evidenced by the multiple copies seen tumbling out of the laserwriter in the general office and lecturers, both male and female, now type up their own student reports, committee minutes, letters and memos, once previously the function of the departmental secretary.

The move towards the academic staff taking over the function of the secretary has not just affected the female lecturing staff, it has also meant that male lecturers have additionally had to learn word processing and now type their own administrative material. As Spender notes, we all have had to make 'the shift to the computer' and this certainly must represent an alteration of the gendered work relations. Green, Owen and Pain also note that 'word processing, the activity and the technology, is shifting in its domain, and the boundaries of the old gendered division of labour (in the office) are being redrawn'.[30] Men have had to learn to type their own correspondance as they are no longer able to draw on the services of a (female) secretary. The old reliance of male staff on the female secretary has changed and has almost disappeared and with it the gendered power relationship of men over women prevalent in traditional office work. However, although men use their computers for word processing it is not their primary activity or purpose as it would be for secretaries or typists. 'Technologies emerge in particular social contexts in response to particular problems, influencing these and themselves, and are reshaped in response to the changing patterns of work organisation and social relations within which they are situated.'[31]

Creativity and Pleasure

'Computers would not be the culturally powerful objects they are turning out to be if people were not falling in love with their machines and the ideas the machines carry.'[32] Many of my respondents have indicated both a need of their computer in terms of supporting their work but also a love of their computer as a close friend and they have indicated this in a number of ways. R10 adores her computer, it has made her into a writer. R11 has developed a personal relationship with her computer, it has become her silicon other, and the power book belonging to R4 is 'friendly', she takes it everywhere with her, while R3 finds her computer companionable with a mind of its own, it is now so much a part of her life that she can't imagine working without it, it has, as Turkle says, 'become an object that they think with' as part of their everyday lives.[33] These quotes are strong indications of the strength of feeling these women have for their computer and how much an important part of their lives it has become and it relates directly to the issues around kinship which I raised in an earlier section.

According to Turkle though, the idea that 'the Macintosh was a friend you could talk to' was a myth generated in the the 1980s at a time when the division between the two computer cultures, Mac and IBM and their clones, was at its height. The IBM PC symbolised a modernist ethos of rationality, logic and accessibility whilst the Macintosh presented its surface to the world to 'play' on. This division though, became 'curiously entwined' with the introduction of the Microsoft Windows software programme which, increasingly, has adopted the look and feel of the Macintosh interface. 'The Macintosh was not just a 'happier' experience for people who were personally comfortable with layering and simulation: It was a consumer object that made people more comfortable with a new way of knowing', though Turkle goes on to say that although its 'cultural presence has increased in recent years' not everyone who uses it is comfortable with it.[34]

On the other hand, Spender believes that men and women have different attitudes towards the computer and states that 'women use them, whereas men fall in love with them',[35] a point not borne out by the quotes above. Women, she stresses, are far more concerned at accomplishing specific tasks on their computers, at getting it to work for them in the limited time that they have available. She considers that men, on the other hand, pursue their interest in their computer as another leisure activity.

The computer has changed the craft of writing for many of my respondents, and, in some cases, this has made them computer dependent, but it has also enabled them to write in ways that previously were either very laborious or impossible. Because of her computer R4 is now able to think differently about her writing, it has opened up her academic creative writing in terms of her use of syntax and grammar. For R1 there is a creativity in trying to make her computer do things not really conceived of by the programmers or software designers, which often necessitates working at the limits of the available technology. Some women though missed the satisfaction that they used to find in drawing now that so much of their time had been given over to writing and were disappointed that the limited tools of screen and mouse were not able, at present, to duplicate this experience for them.

But although many of the women 'loved' their computer, the transition from print to computer has not always been an easy one. Our sense of self has had many knocks along the way in the acquisition of basic computing literacy and for some the frustration of trying to learn was only relieved when they acquired a Mac. Now we recognise realistically that, however hard we try, with whatever time is available to us, some of us will never really be able to keep up with it all.

Conclusion

In writing this paper I have attempted to integrate the personal, anecdotal recollections of my respondents and their relationship with a computer, with the broader cultural shifts that have occurred in the world in which they work, the University, and I recognise the limitations in my approach. I have started from the broad premise of the University as a generic institution and a 'known' site and have not made any reference to the myriad other changes that have occurred within Universities. Modularisation, expansion in student numbers, declining units of resource, and the extension of administrative duties have all impacted on the working lives of my respondents, the students and indeed the structures and functions of the institutions themselves.

My case study has also been restricted to a small group of middle class, professional women which, should I wish pursue this initial research any further, would need to be extended to include a wider spectrum of women, in respect of age, ethnic origin, class and status within their place of employment. In some ways though restricting my respondent group has been an advantage as well as a disadvantage. As I mentioned in my introduction, the women who responded to my questionnaire possess a strong group identity, some of them belong to the research group 'Cutting Edge' that have been involved in editing this book, and all the respondents are personally known to me. This has meant that I have been able to draw on a set of shared experiences and language values in my reading of the texts. It has also meant that I have been very subjective in my interpretation of their responses. My familiarity with the respondents has perhaps meant that I have made too many assumptions about what I thought they were saying.

All the women have had to use a computer as an essential part of their work and all have experienced, in similar ways, the shift from print to computer based applications in their place of work, more or less in the same sort of time period, the late 1980s and have, in the main, received this change positively. It has been interesting, therefore, to identify how selective the respondents have been in assessing which aspects of their relationship with computers have been of value and which they have rejected. Ironically, it seems to me, one of the biggest drawbacks must be the dependency expressed in relation to the object of their growing 'affection', the computer, and the sense of loss they would feel if they no longer had access to one. Most respondents stated, emphatically, that they now had to wait to start writing until they could do so on their computer and that they had gradually lost their facility for handwriting except for 'special' or 'private' occasions. Also, they compared the amount of time that is now taken up in tuning and fine tuning word processed text, with the laborious experience of writing and rewriting texts out by hand.

So what of the benefits? How has the computer been useful in enabling my respondent group of women to experience positively, their sense of selves? Learning to use a computer has not always been an easy experience and it has taken time, at least, for some of them to be able to take control of the technology and make it work for them. Access to the technology, specifically the computer, has also been a struggle both in terms of staff development, learning the software, and availability of the machines to gain skills and expertise. Some of the respondents now work with computers as part of their art practice, they can get the technology to work for them, to a degree, and will commission programming specialists when they need more specific expertise. None of the respondents has altered her computer either by hacking into the computer's internal system or by reprogramming the software to make it 'do' individual tasks. 'In

182

matters of technological change, women are more impacted upon than impacting'[36] and therefore in order to be able to effect change in the way the technology works for them and their access to it women need to become more involved in its use and to make more demands that the technology is designed and performs in a way more suited to their needs.

To go back to Cockburn, whom I quoted at the beginning of this chapter, '... contemporary Western femininity has involved the constitution of identities organised around technological incompetence'[37] and it has therefore been rewarding to hear my respondents refute this claim and talk instead with pride of their ability to take control of, and become confident with, communications technologies.

Sangre Boliviana:
Using multimedia to tell personal stories

Lucia Grossberger-Morales

My Personal History

I emigrated from Bolivia to the United States when I was only three, yet it has been one of the most profound experiences of my life. On my fifth birthday I swore that I would never forget the pain of being taken away from my mountain home, from my extended family and from my language. On that birthday I knew that one day I would find a way to tell my story. I had moved from the third world to the first world, from a rural town to one of the world's largest cities. I also left an extended family, where I had felt confident and safe. Emigrating to New York made me and my family feel alien and helpless. I felt despondent because few understood my Spanish words and in this new land, I felt I had lost my voice.

This was very distressing to me as I consider the use of Language to be very important in the way I think about and approach my art. When I switch from Spanish to English, I am also switching cultures. As a cultural hybrid, I am inspired by Latin American art, literature and culture and by personal computer technology. I am an artist and the goal of my work is to weave the two cultures, Latin American and North American, together, on a computer, in a way which is unique.

The work of Eduardo Galeano has been an inspiration to me and has been central to the writing of my own personal history. Quoting from the Preface of *Memory of Fire: Genesis*, Galeano writes:

> '*I did not want to write an objective work – neither wanted to nor could. There is nothing neutral about this historical narration. Unable to distance myself, I take sides: I confess it, I am not sorry. However, each fragment of this huge mosaic is based on a solid documentary foundation.*

What is told here has happened, although I tell it in my style and manner.[1]

Learning about Computers

In 1979 when I first got my Apple II+ computer I knew I wanted to use this new technology to tell my stories. For many years I had experimented with film, video, and animation to tell stories; these were imagined dream-like stories that looked like animated paintings, which had evocative voice-overs. I had found that traditional animation, although seemingly the most likely medium, was just too cumbersome and in 1979 I bought my first computer, a primitive Apple II+ personal computer. I knew that there was enormous potential for this medium, it was becoming less expensive and more powerful in leaps and bounds and I felt that it would come to play a very important role in society. Prior to my buying my first computer I had been completely awestruck by a paint program, AVA, developed by Ampex, this was the computer system, costing a quarter of million dollars, which made me realise that the new computer art was the medium that I wanted to work with.

Initially, it was very difficult for me to produce my art with this new medium. The hardware was chunky and primitive. The Apple II+ high resolution mode was 140 by 192 pixels with four colours, purple, green, orange and blue, and there were limitations on what colour could be placed on what x, y coordinate. There were few software packages available and those that were had been designed by hackers who assumed the user would have a knowledge of programming. I was also not able to take any computer graphics classes because I didn't have the necessary math requirements. But there were some reasonably simple books on learning the BASIC programming language and I discovered that the computer is an endlessly patient partner and so, gradually, began to understand the Apple II+ hardware.

In 1980, with two collaborators, I developed the 'Designer's Toolkit', a paint program that was published two years later, by Apple Computer, Inc. It was widely acclaimed as one of the best graphics programs for the Apple II computer and it was one of the first programs that was designed by artists, for artists. I developed several other book/software packages for the Apple II between 1980 and 1987.

185

But by then I'd had enough of designing and just wanted to create my own art, and to tell my own stories.

In 1982, I had created my first computer installation at a cultural centre in the Silicon Valley. It included an Apple computer and 'The Designer's Toolkit'. It was such a new technology that a full time attendant was required to guide the viewers on how to use it. Things have drastically changed in the last eighteen years; now my biggest concern is to make my installations crash proof and to construct a multilevelled piece so that a less experienced user can still enjoy it and a more experienced user would feel challenged. I also aim to use the least amount of text possible to indicate how my piece should work, at the most, one sheet which gives simple navigation instructions.

In 1984, the Macintosh was introduced and it changed the world of personal computers so much that, over a ten year period, I could do everything I wanted and more than the original AVA system I had admired in 1979. Now, with a Macintosh computer, the storytelling potential of multimedia seemed unlimited.

My Way of Working

I have been working as an artist using various multimedia techniques on a personal computer for nearly twenty years and have worked on games as well as educational software.

I have identified the following goals that I used to help me with my work on 'Sangre Boliviana' and 'A Mi Abuelita' and have listed them here as:

1) The need to tell my personal history in my own (bilingual) voice.
2) The ability to take advantage of every aspect of multimedia technology: digitisation and integration of moving and still images, video, sound and words.
3) The priority to design the content of each piece so that it should match its interface. This is one of the real challenges for me and part of the process on which I probably spend most of my time.
4) The exploration of the computer's ability as a procedural tool to follow a set of directions or programmes.
5) The ability of my work to be as interactive and participatory as possible.

6) The possibility of making my work available to as many people as possible.

Before I started working with computers I used to be a painter and so I have tried, in some of my work, to recreate the feel of paintings. For instance, in 'The Dream', and in 'Catavi '89', sections from Sangre Boliviana, I have used various filters in Photoshop, Paint Alchemy and Painter to manipulate my photographs so that they would look like paintings. In 'The Dream' the final piece looks like a series of animated impressionist paintings and in 'Catavi '89' I created processed, surrealist images that reflected my feelings of seeing my birthplace which had become a ghost town.

I often show my work in museums or in museum-like settings and therefore like to create an environment in which to display them, as in the case of 'A Mi Abuelita' and 'Sangre Boliviana' where I have made installations like shrines.

In 'Sangre Boliviana' I have also used a combination of different sounds, music from the Andes, played on traditional Andean instruments, and voices, sometimes together and sometimes on their own. I also use written text to say one thing while the voice over text would say something different. I use the voice to speak the language of the heart, while the text is used to present information.

I am also concerned that using text on a computer is very different from reading a book or a magazine and believe that these differences need to be taken into consideration when designing text for multimedia. Long blocks of text with scroll bars are difficult to read and navigate through. The screen resolution for text is not ideal for reading. The screen monitor is often too wide, producing line lengths which are difficult to read. Graphic designers have come up with some good solutions to these problems, but I believe that writing for the screen, when it is intended to be read off the screen, is a unique situation which needs to be addressed with a different style of writing. I have attempted to do this by using language sparsely, much like I would work on poetry.

For example the text for 'A Mi Abuelita' is the following: 'My Grandmother Modesta was born in 1879. She spent her whole life in Cochabamba Bolivia. She married Luis Zambrana Mendoza. They had

Wednesday I went to her bedside.

She told me the angels were coming to take her.

Figure 1
'A Mi Abuelita'
screen shot,
1992 - 1994.

fourteen children. In 1969 I visited her. There was Abuelita, savouring life, drinking Chica. Then on a Tuesday she got ill. On Wednesday she told me that the angels were coming to take her. On Thursday she died.' Added to the animation and images the simple text tells a powerful story.

The fact that I can create such complex pieces that can contain music, video, text, animations which are interactive and which can be inexpensively reproduced, is, to me, very exciting. Maybe it will be computer art that will redefine the art world and rather than valuing exclusivity people will value what is meaningful and can be shared. Additionally, since the work I do is digital and, unlike video, the copy is identical to the original with no loss or degradation in quality, the purchaser is free to copy the images or code that created the original piece. I have made a policy of including the code that created my work on my CDROM's because when people see my work, they become excited about telling their own stories and I want them to understand how they could possibly do that, using my code and the increasingly inexpensive new media computing technologies. I am completely aware that my views are not in tune with the attitudes of the copyright laws and that most people feel proprietorial about their work. But for me, sharing the process of making the work is as important as the work itself.

The Internet, though currently very limited in its graphic capabilities, also has tremendous potential in regard to reaching a mass non-elitist audience. A very abbreviated form of 'Sangre Boliviana' is available at: http://tmn.com:/Artswire/Sangre/SangreTrilogy.html. Additionally, for one of my museum installations, by clicking on the word 'Web' from the main menu in 'Sangre Boliviana', Netscape Navigator was launched and on the bookmarks was a list of other sites that dealt with Bolivia. The idea of combining the CDROM with the Internet therefore also holds some very exciting possibilities.

Two Interactive Multimedia Installations

In this section, I will describe the two interactive multimedia installations that I have created which explore language and culture, 'A Mi Abuelita' which translated means 'To my Grandmother' and 'Sangre Boliviana' which means 'Bolivian Blood'. Both installations tell stories which have their basis in Bolivian history as well as my own history and I have provided links so that the user can navigate and experience these stories in many different ways. All the stories in 'A Mi Abuelita' and 'Sangre Boliviana' have evolved and changed over time as I have integrated new ideas, developed my skills as a painter and extended my potential for using the software.

'A Mi Abuelita' (To My Grandmother)

I created 'A Mi Abuelita' in 1989, as a shrine dedicated to my Great Grandmother, Modesta, for *Dia de los Muertos*. Day of the Dead is a holiday in many parts of Latin America which honours and celebrates the dead. The family gathers in the graveyard, creates a shrine, prepares the favourite meal of the dead and drinks 'chica', the popular corn beer. A plate of food is left out for the dead, so they can share in the meal.

There are two parts to the installation 'A Mi Abuelita'. The first is a shrine (Fig 2, page 162) which takes the form of an animated story (in Spanish and English) of my great grandmother's life. The second is also a shrine (Fig 3, page 163), this time a computer installation

189

that confronts the viewer with different aspects of my grandmother's death. In Latin America children are given toys that deal with death and death plays a large part in their culture, it is the other side of life, it is one of the same duality. To get into the second part the viewer has to go (literally) behind a curtain to view the computer shrine where, as they get closer to the shrine, they see themselves on a monitor which suddenly triggers an animation of a skull that disintegrates. I wanted this to shock the viewer as well as to amuse them. On this second shrine there are also several artefacts from Latin America; a bowl containing 'mummies', which are used instead of skulls for the Dia de los Muertos, baskets of local delicacies, typical and specific to the Andes, 'Quinoa', the mother grain of the Andes and 'Chica' the traditional beer, made from corn.

There are several links that enable one to navigate in different ways through the story which are indicated by italicised text. For example, clicking on the italicised word 'Chica' brings up the section on the cultural values and history of Chica.

These links out of the main story develop the context of my great grandmother's life, without taking away from the elegance and simplicity of the main thread. This is a story telling technique, a system of navigation, which is not totally new to the medium, but which is a particularly elegant solution to presenting a very simple, childlike story and being able to integrate within it the different links that include information that only some readers would be interested in.

'Sangre Boliviana' (Bolivian Blood)

My second installation, 'Sangre Boliviana' is a CDROM, which includes an Internet connection. Started in 1992 it explores my personal relationship with Bolivia, the country where I was born. Bolivia is a small landlocked country of only six million people. Seventy percent of the people are indigenous speaking their native languages as their first language, their second language is Spanish. In the Bolivian Andes, one can drive from the snowy peaks to a jungle with orchids growing in a matter of a few hours.

This is my homeland which I left at age three for America, and I return to my roots to weave my stories, woven from my history and Bolivia's history and culture.

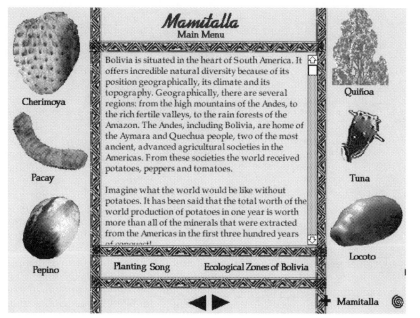

Figure 4
'Mamitalla',
from 'Sangre
Boliviana', 1996.

'Sangre Boliviana' is a postmodern collage of different fragments of this history. I have woven it together using still images, sound, video, voice, clip art and drawings integrated with the help of various software applications onto Director. It contains eight interactive multimedia pieces which explore my personal relationship with Bolivia. It has been inspired by Bolivia, and the compelling beauty of the Altiplano, (the high plain) and its indigenous people. Each of the eight pieces displayed on the main menu have at least one branch to other sections. Each section has its own style and its own interactive story told through words, images, animation, video and sound. One of these pieces 'Mamitalla' focuses on my relationship to the Andes and its indigenous vegetables and fruits. It is not only about the diversity of Andean crops, but also about the Inca cosmology and its relationship to Mother Earth. At the time of the Spanish conquest in 1533 Inca agriculture was highly developed, producing an abundance of crops almost doubling that of Europe and Asia combined. The Andes has provided the rest of the world with potatoes, tomatoes and peppers.

Figure 5
'Palabritas',
(Little Words)
is an interactive
picture/poem
from 'Sangre
Boiviana', 1993.

'Palabritas' is an interactive picture/poem from 'Sangre Boliviana', 1993 which includes a Quicktime movie. The interface enables you to playfully move from Spanish to English by dragging the cursor over the text, when you click on the doll the piece 'Muñeca' appears. 'Muñeca' is a visual environment that branches from 'Palabritas'. It is a room containing strange objects. When you roll your cursor over the sun a strange doll will appear, roll over the window and you will see a dog. Click on the artichoke and this will enable you to read a story about the doll.

All eight pieces in 'Sangre Boliviana' tell my story of being an immigrant in America and of growing up in two cultures.

Creating 'Sangre Boliviana' was an organic process. I would get an inspiration for a section, it might be a story or dream, or it might be a festival or ritual. I would then piece it together and let the section dictate the interactive format. I made no attempt to follow a time line or overall design. My goal was to follow my intuition as much as possible. I have been working on 'Sangre Boliviana' for over five years. Much of that time has been spent on matching the content and the format with my initial inspiration for the sections, and,

Figure 6
'Muñeca', (Doll)
is a visual
environment
that branches
from 'Palabritas',
part of 'Sangre
Boliviana', 1993.

because working with multimedia is such a fluid medium and the possibilities for changes are seemingly endless most of these sections have gone through several iterations.

Conclusion

Telling stories is a vital part of what makes us human. The more I have shown my work, the more I realise that much of my work is not really about my stories, but rather about inspiring others to tell their stories, using this new medium, multimedia. Interactive multimedia enables us tell more complicated stories, it allows for branching and layering, and telling the same story from a variety of different perspectives, creating sounds and visual images that would be virtually impossible without this new technology. Interactive multimedia is for me, without a doubt, the medium that best allows me to best express and relate to others, what I see, and hear, in my imagination.

Disappearing Digitally?
Gender in the digital domain

Jackie Hatfield

Introduction

In researching for my PhD (the practice component, which involves developing interactive moving image installations using programming), I had started to look once more at Dale Spender's study of language in *Man Made Language* which was first published in 1980. Until that point I was primarily interested in the construction of imagistic language (film and video) specifically in relation to gender representation. I was concerned with the fact that historically there has been a gender bias in favour of the masculine/male. Certainly even in the 20th century women's proactive involvement with mass cultural production has been sparsely recorded and discourse around the signification of the main forms of production, i.e. the image and the word, is relatively recent. Considering Dale Spender's arguments that women's exclusion from languages is evident through the construction, and Luce Irigaray's arguments that exclusion takes place inherently throughout structures that are dominated by the phallocentric, I was concerned that the hyperbole that computer technology has the potential to empower women may be short-sighted.[1]

> *Thus instead of remaining a different gender, the feminine has become in our languages, the non-masculine, that is to say a non existent reality.[2]*

In this chapter I am primarily addressing the language of computer technology and will consider how coding may be the singular and most significant element in the future production of meaning. I will demonstrate the logic of my argument by looking at parallels with the printing press and the effects that this major technological revolution had on women. Similar to the computer the printing press provided a conduit for the mass communication of ideas through a language to a populace, though through this system of industrial production women's written contributions were marginal relative to their male counterparts.[3] The societal reasons for this are varied, including limited access to education and limited economic means etc. but it is the issue about the primacy of literacy as a means of communication and women's access to this aspect of means of production that is the link to my argument about computing.

In a sense, in considering this historical model, both in terms of instrument and language, I am pre-empting a possible parallel in terms of women's visibility between the printing press and the computer. The printing press was the first technological transmitter through which one form of language – the word – took dominion. What kind of access did women have to this invention? My argument is that as women we should be careful not to underestimate the significance to our gender of the proliferation of meanings defined and shaped through the software of computer technology. Indeed, we should consider the legacy it may leave in terms of our own exclusion from the symbolic.

The computer is not merely a tool or a mechanical device, like a car for example. Computer software, which is part of the architecture of the machine, presents us with an illusion that we can expand our symbolic and imaginative spaces by simply learning how to use it. The potential of what software can do is limited by the commercial structures that dictate perimeters. Market forces determine the production of software and currently men dominate computer-coding authorship.

Disappearing Digitally?

Computer software is made up of the convergence of three specific languages; the word, the hieroglyph, and the digit. In this chapter, I will be considering two aspects of language, the word and the digit.

The knowledge of language/s is a determining factor in what is being said and by whom. Looking at the effect technologies have had in the way Western gender identities have been represented in language, does this have any significance for how we (as fe/males/ wo/men) should be approaching the languages of the computer?

195

Programs and software are developed with sales in mind. They are the front end of the computer/user interface. In *Life on the Screen, Identity in the Age of the Internet,*[3] Sherry Turkle describes how computer interfaces have developed from being relatively transparent to opaque based around the windows style design of the Macintosh interface. In the 70s and 80s, early personal computers were developed as systems that encouraged the user 'to think of understanding as looking beyond the magic to the mechanism'.[4] In contrast to this, the introduction of the Macintosh in 1984 provided an interface that 'did nothing to suggest how their underlying structure could be known'.[5] Turkle's study identifies different types of user relationships to computers, but she argues that there are differences in the user relationship between the IBM PCs of the 80s and the contrasting Macintosh.

> *Post-modern theorists have suggested that the search for depth and mechanism is futile, and that it is more realistic to explore the world of shifting surfaces than to embark on a search for origins and structure. Culturally the Macintosh has served as a carrier object for such ideas.*[6]

Turkle suggests that it takes a particular type of user to get into the machine and go beyond the surface interface. Hackers and what she describes as 'hobbyists' are users who aim to understand the workings of the computer and make it transparent, not only taking the hardware apart, but understanding the construction of the layers of software. Coding or programming is a form of language, it is not only a tool, it is partly the nature of the thing, and related to the way the hardware functions. 'Underlying all forms and roles of software is the language we use to communicate with computers.'[7] The issue is that the user is engaging with many layers of computer coding language when using software, but at a surface level. What are the consequences? In terms of the creativity and authorship of image manipulation for example, the creative perimeters of software are already pre-defined by a programmer. The artist/author who uses the software is working within a coded imagistic language that is prescribed by someone else therefore authorship of the artist is relative to the already written creative permutations by the coder.

Like the fundamental shift in the mass dissemination of ideas in Europe that happened with the invention of the printing press, the computer and the Internet have very quickly shifted our perception of the potential of communication. Like the letters that were the integral part of a printing press, without which the instrument wouldn't be functional for the dissemination of meanings, so the computer will not function without the language of coding. What that coding is, and the functions it is compiled to do, are determined by whoever is literate in that language. Whoever is not literate will be excluded from the activity of 'reading' and 'writing' it. Historically through the construction of the word language, women have been excluded and our nature has been assumed. The French Feminist Luce Irigaray argues that sexual difference 'conditions language and is conditioned by it. It not only determines the system of pronouns, possessive adjectives, but also the gender of words and their division into grammatical classes: animate/inanimate, concrete/abstract, masculine/feminine, for example'.[8] She goes on to argue:

> *Just as an actual woman is often confined to the sexual domain in the strict sense of the term, so the feminine grammatical gender itself is made to disappear as subjective expression, and vocabulary associated with women often consists of slightly denigrating, if not insulting, terms which define her as an object in relation to the male subject.*[9]

In relation to the computer, we cannot expand the potential of the construction of meaning through coding if we don't know how these new languages are constructed. Considering the front end of software without question is the equivalent of reading a book and not questioning the exclusion of 's/he', or watching a film and not questioning why the transgressive 'wo/man' always gets 'wasted' or punished. As

197

women we need to be literate/numerate in at least some basic language/s of the computer, beyond the surface interface, so that we can define the functions of the instrument for ourselves.

> *Language is a human product, it is something which human beings have made, and which can be modified. We can – with perseverance – posit alternatives to those, which are readily available within our society. We can make the effort to formulate possibilities at the periphery of our cultural conditioning and to reconceptualize our reality: we can generate new meanings – and we can validate them.*[10]

Computers and the Internet, as places of congregation, are cited through the media and also in academic works, as revolutionary developments that will enable the freedom of the individual, and be a positive development for the socialisation of groups that were previously excluded or marginalised. The reality is that the majority of the global population has no access to a telephone, let alone a computer or the Internet. For example, Sadie Plant likens the development of the Internet to the telephone, arguing that it is simply another means of communicating:

> *What was supposed to be a simple device for the improvement of commercial interaction has become an intimate chat line for both women and the men who once despised such talk. And as means of communication continue to converge, the Net takes these tendencies to new extremes.*[11]

Although the Internet *is* a development of telephony, there are other issues beyond simply communication and exchange. The fact is that it is relatively expensive to use the internet, and computers are not being bought by all social groups. In the US there is a 'digital divide' that has increased between certain groups between 1994 and 1997.[12] In a country that has a mass telecommunications policy 'that all Americans should have access to an affordable telephone service',[13] among those least connected (to the Internet) via modem or PC are identified by the US Bureau of the Census, US Dept of Commerce as: – Rural Poor; Rural and Central City Minorities; Young Households and Female-headed Households.

> *Single-parent, female households also lag significantly behind the national average. They trail the telephone rate for married couples with children by ten percentage points (86.3% versus 96%). They are also significantly less likely than dual-parent households to have a PC (25% versus 57.2%) or to have online access (9.2% versus 29.4%). Female-headed households in central cities are particularly unlikely to own PCs or have online access (20.2%, 6.4%), compared to dual-parent households (52%, 27.3% or even male-headed households (28%, 11.2%) in the same areas.*[14]

It is the optimism around the democratisation of technology that I am arguing against, and the premise that the computer is simply a conduit, a tool, because it is actually a system that involves complex language structures, and inevitably social structures that revolve around the machine. Nicolas Negroponte, director of MIT and founder of *Wired* magazine, concludes his book *Being Digital* by stating 'Like a force of nature, the digital age cannot be denied or stopped. It has four very powerful qualities that will result in its ultimate triumph: decentralising, globalising, harmonising, and empowering'.[15] Negroponte disassociates the digital domain from any external controlling force, arguing that it will be a world populated by decentralised individuals who will alter and shape a 'greater world harmony'.[16] But the people who inhabit digital space are no different from the people who inhabit the real world and in terms of gender equality, the same prejudices apply. The Internet or computers in themselves won't change ideologies about gender. As Sadie Plant clearly states about the telephone, it was '... intended simply as a means of conversing at a distance, and not designed to redesign talk itself ...'.[17] In her book *The Pearly Gates of Cyberspace: A History of Space from Dante to the Internet,* Margaret Wertheim recognises that

199

cyberspace would not exist without the collective agreement of the electronic community to use compatible electronic languages. 'Cyberspace as a communally shared world would simply not be possible without the immaterial power of language.'[18] She goes on to discuss the production of electronic languages and again recognises the significance of computer scientists, language developers to the development of cyberspace.

In *Media Technology and Society* Brian Winston argues that the promotion and glorification of the potential of the 'information revolution' implies that there is currently a sense of amnesia in the understanding of the impact on society of previous technological achievements. Winston's argument is oriented predominantly around hardware though his approach is particularly pertinent when his basic premise is applied to female involvement with technology and whether women can consider the current change in technology a revolution, or indeed a positive step in terms of women's emancipation. Understanding the impact of previous 'technological revolutions' on gender should give us an insight into whether the current 'information revolution' will truly enable women or whether the hegemonic patterns that exist in real space will be paralleled in virtual space.

Words

In a short space of time of around thirty years during the latter part of the fifteenth century the printing press was introduced all over Europe and created an instrument for the mass dissemination of ideas on an unprecedented scale.[19] Hundreds of printers worked across Europe, often transporting their printing equipment from one place to another. Significantly, printers developed a tradition of marking their authorship by placing a personal mark on each book. This was primarily for commercial reasons to denote ownership or origin, in effect the books became commodities.[20] 'In adopting the idea of the trademark, therefore, the early printers recognised that the books that they produced were different from the books made by scribes. Printed books were thus early identified as articles of commerce, moving in the channels of trade.'[21] The conjunction of the technological invention with commercial interests created a context for the proliferation of certain stories or genres, in the fifteenth century almost half of the books printed were of a religious nature, biblical or theological. Contemporary writers accounted for around fifteen per cent of books printed.

> *To some degree, exclusion from the canon results from women writers' being obscured rather than scorned: there is safety in numbers, and undoubtedly there were more men than women writing in the Middle Ages.*[22]

That the majority of authors and printers were in fact men would have meant that their own gender pro-activity through the language narrative and symbolic space remained unchallenged.[23] If there was sexism in language before the printing press its subsequent mass dissemination was guaranteed. To a large extent, it was the social economic context that rendered this industrial form of technology inaccessible to women, and mainly their exclusion from education that left them illiterate and therefore unable to challenge the new dominant form of communication.[24] It has taken five hundred years for women to begin to address the gender inequality that is manifested in words. In the twentieth century it is through education and independence that women have learnt to be distrustful of phallocentric pre-determined codes and languages. Through historical precedents we have learned to understand how our gender has been marginalised and excluded from language and we have challenged this exclusion.

> *One of the basic principles of feminism is that society has been constructed with a bias which favours males; one of the basic principles of feminists who are concerned with language is that this bias can be located in the language.*[25]

Whether we agree or not that one form of feminism can be a term for all feminisms, Dale Spender's arguments that sexism is apparent in language still stands. In *Man Made Language,* Spender makes reference to the gender and racial bias evident through the structure of linguistics, both in semantics and syntax.[26] Spender argues that through the construction of language women have less status than men, and that this is visible both through the derivations of individual words, and in the way that sentences and grammar are compiled. If women do not have equal status with men in the linguistic field, then it is safe to assume that this reflects similar gender inequalities within society in general.[27]

> *This social and cultural injustice, which nowadays goes unrecognised, must be interpreted and modified so as to liberate our subjective potential in systems of exchange, in the means of communication and creation.*[28]

In her book *Thinking the Difference* Luce Irigaray proposes a system of solutions for women in a culture that she argues is dominated through language, law, art, science and technology by phallocentric codes and structures. One of those solutions is to address the mythologies perpetuated through discourses that have excluded particular female oriented narratives. Irigaray argues for example that 'father-son, mother-son relationships dominate our religious models'[29] and that there is no representation of mother-daughter couples (apart from in advertising) in religious or civil sites. It is one visible aspect of gender inequality and, Irigaray argues, a consequence being 'Women's inability to organise themselves and agree on what they want makes some people smile and discourages others. But how could they unite when they have no representation, no example, of such an alliance?'[30] In her discussion around the gendering of words, Irigaray says, 'One way or another, the gender of words is related to the question of the gender of the speaking subjects. Words have a sort of hidden sex, and this sex is unequally valued depending on whether it is masculine or feminine.'[31] Irigaray is arguing that gender inequalities diffuse into every aspect of women's daily lives, from how we use words to describe our identities to the construction of social systems such as the law and religion.

If language reflects inequalities then the mass dissemination of language only compounds them, in effect inequality becomes a fait accompli, unequivocal and difficult to challenge. Linguistics change slowly, and although women have made strides in understanding the 'construction' of their silence when it comes to the word, what about the primary language of construction of the computer, the code?

Digits and Wo/men Computers [32]

Ada Byron Lovelace 1815-1852 was the first female computer programmer and she is discussed at length by Sadie Plant in her book *Zeros and Ones Digital Women + the New Technoculture.* For the purpose of my own argument Lovelace is crucial to the history of computing because she embraced the language of computers – mathematics – and determined to understand the construction of the instrument. Her interest in mathematics cannot be underestimated for it was this that led her into her involvement with the development of Babbage's mechanical computational machine – The Analytical Engine.[33]

All the women who were crucial to the development of computing were specialised mathematicians.[34] There were hundreds of other unnamed women 'computers' who were involved with code breaking in the 1940s during the Second World War at Bletchley Park in the UK, and the University of Pennsylvania's Moore School of Electrical Engineering in the US. Computational skill was crucial to the work of code breaking and to the development of computational hardware.[35] At the University of Pennsylvania, it was women with mathematics degrees who were assigned to this national task.

In light of the fact that currently so few women enter the programming profession it is interesting that it was Grace Murray Hopper who developed the first computer compiler in 1952. She initiated the development of a comprehensible and universal computing *language.*[36] Hopper was particularly interested in developing a programming language that was accessible and easy to use rather than the complex binary system that had to be compiled each time a code was written. Her research was integral to the development of COBAL,[37] the first near universally common business language developed collectively in 1959 that would run on any computer. Her software development was one of the first outside the pure zeros and ones of the machine language. By using computational mathematics as a source to devise codes and programmes for computing, Grace Hopper's research was integral to the development of computing languages. Highlighting the fact that Grace Hopper was a woman and not a man is only significant to illustrate the point that there were a number of women who were integral to the development of computing and who do access this profession, but not nearly enough. Education is already gender determined and there is a lack of emphasis on mathematics and construction/de-construction skills in education for girls.[38] The issue is important when considering the creative possibilities of coding in the area of software development, which relies on computational skill and the potential for girls and women in this field.[39] According to Sherry Turkle the programming methods she experienced in her 1978 programming class at Harvard still prevail. 'Where my professor saw the necessary hegemony of a single correct style, I found a range of effective yet diverse styles among both novices and experts.'[40] Turkle goes on to discuss the structured method of programming that is still seen as the 'definitive procedure' by many computer scientists.[41] Students who tried to pursue diverse approaches to programming were in effect alienated. It is possible to deduce from Sherry Turkle's experience and the decline in the number of female computer graduates that the patriarchal order Irigaray argues underlies structures within our society is alive and well and perpetuated through the education of programming.[42]

Mathematics as Language

> *Virtually all software today still retains its heritage of mathematical logic. Current computer software design still uses, almost exclusively, functions grounded in mathematical logic.*[43]

On Saturday February 20th 1999, the *Guardian* reported on a survey by 'The Centre for Economic Performance' that 'The secret of getting on in life is not literary flamboyance, as arts graduates often insouciantly assume, but a familiarity with the more recherché applications of calculus'. Anna Vignoles who carried out the research at the University of Newcastle came to the conclusion that graduates with Maths A level earn more that those without. This was reported by the *Guardian* as being the result of a shortage of highly numerate candidates in the job market. 'In some respects, the education system is currently failing employers.'[44] The statistics are that only 9% of all 1997 'A' Level entries were in maths. This research did not account for gender bias.

In the subject area of mathematics the gender bias in secondary education is weighted in favour of males. 'By the sixth form, a pattern is well established of arts subjects (especially languages) being predominantly girls' subjects, and maths and sciences (other than biology) being predominantly boys' subjects'.[45] (Incidentally it is interesting that maths is not seen here as being a language.) The Department of Education and Science statistics for 'A' level passes in maths, physics and chemistry for 1982 were 72,000 boys to 29,000 girls; Department For Education statistics for 1993 were 51,000 boys to 26,000 girls. If we apply Anna Vignoles' argument, these statistics can be translated into the proportion of women who enter computing science professions. In fact in the US, and more so in the UK, the numbers of female computer science graduates have declined since 1984.[46] If girls and women don't gravitate towards mathematics as a subject area during their primary and secondary education and importantly at 'A' Level, then they aren't as likely to go into areas of employment that employ mathematics such as software development. If women don't work on the development of software, that is, the language of the computer, then where does that leave them in terms of literacy in this field? If girls and women aren't taught that mathematics is crucial to the development of computing language and also to their involvement with writing software, how can they start to be active in this area? It is ironic that the impetus for user friendly computer language was initiated by a woman, and certainly in the light of the progress women had made into computing during the Second World War that women are currently so poorly represented in the computer coding professions.

Mathematics is a language that is translatable into many areas of work and I would argue that the application and significance of mathematical formulae and calculation in relation to computing is crucial to the development of the software. Computing languages are formed by using mathematical and computational processes, the history of the computer runs hand in hand with computing or mathematical calculations. In the twentieth century it is significant that women have fought to understand where their own contribution lies in relation to the languages of the image and the word, and to understand their formulation. Crucial to women's visibility within those languages is their ability to subvert stereotypes and tell their own stories. What about coding and the development of software? What about women's understanding of the language that is universal to these? The questions to be asked are: are the computer languages that construct software phallocentrically dominated? Do they inherently exclude women, as Dale Spender has argued about language and as Luce Irigary has suggested about all other structures in society?

Software is encrypted so that its coding source and construction cannot be accessed which makes it difficult to read 'manifestations of a patriarchal order' from computer coding.[47] Dale Spender cites examples of gender bias in the semantics of linguistics, but it is difficult to determine coding equivalents of this. Another way to determine the nature of gender bias as well as the nature of women's involvement with computing in the workplace is to consider the question at a different level, that is, the commercial output of software. As I stated at the start of this chapter, market forces (and historically the military) determine the potentiality of code. Currently computer games are marketed predominantly to males, computer software developers assuming that there is a strict gender difference in what is marketable. The prevailing cultural ideologies about gender are manifested in games that in turn proliferate gender stereotypes and gender bias in favour of the male. Looking at current commercial output outside a gender polarisation position and only in relation to market choice, there is an abundance of one type of strategy game (i.e. shoot and kill/destroy/maim and win) and little market choice.[48] Even if you are a girl and you like to play shoot and kill games, marketing is aimed at boys; there is already a prevailing gender ideology here, that it is only boys who like kill games.[49]

Gender inequality in software is engendered through gaming narratives and visible through the availability and marketing of certain gaming narratives over others that predominantly reflect the interests of men. This is a kind of evidence that women are not creating

206

their own symbolic and imaginative space within this aspect of technology. Similarly to the previous historical models I have cited I am aware that the reasons for this are varied, and that the applications for coding are immeasurable. What is important to my argument, and also to women is that they are given opportunities to access the means of production of the computer, i.e. coding, as it is this aspect of the computer that is so crucial to its potentiality. Without women participating in the creation of narratives, our experiences and identities will be marginal via this technology.

> *If present trends continue and girls are not encouraged into maths and science and computer programming, they will be trained only for the data and information-retrieval capabilities of the computer. These are still secretarial/ clerical skills, and females will remain at the low end of the service-oriented pay scale.*[50]

In talking about the computer age, or what tends to be described as the 'information revolution', I am making a distinction between two main areas of development, the hardware (i.e. the computer, the transmitter), and the software (the language/s). The hardware is a conduit; it is the software – the code or the language – which is the significant element in women's future visibility and involvement in making meaning within the domain of computing technology.

It isn't easy to determine aspects of phallocentricity within the machine; in fact, it is problematic. It isn't the purpose of this chapter to be negative about the technology of the computer or women's involvement with it. What I'm aiming to do is to highlight the relationship between mathematics education and programming education, and to stress how the paucity of women involved in these areas may effect gender representation within the digital domain. This is not to negate the women who are forging their identities with the aid of computer technology either through the Internet, through programming or by using existing software. As women we need to go beyond the surface interface of the machine to determine our own set of rules and structures that are not defined by men. We are in a moment of transition with this new instrument and women should recognise that computing languages are *constructed*. The important thing is not to underestimate the potential of coding to enable us to construct meaning. As women, we do not want to be excluded from the construction of the imaginary within the languages of the computer as we have been historically excluded within other forms of language.

Voxpop Puella
A work in progress

Helen Reddington

In 1977, like many young women in Brighton, I joined a punk band; I was excited by the mixture of politics and noise, an abrasiveness that was very pertinent at a time when youth unemployment hit society between the eyes. There seemed to have been no forewarning and there was definitely no national strategy in place to cope with such a quantity of young people, bored, impoverished and restless. With a bass guitar borrowed from Poison Girls (a band whose female lead singer was over 40 years old), and clothing found on the landing of the squat where I lived, I joined a throng of mixed-gender bands who felt that revolution was in their hands.

The movement was short-lived: all new styles of music in Britain, whatever their origin, seem relegated to the realms of fashion. Nationally, we have a short attention span with regard to the arts and music; politically, the advent of Thatcher's entrepreneur-friendly reign divided the ex-punks into those who wanted to capitalise on the fruits of their labours and those whose desire for independence was more important to them than anything else.[1] Countercultural music has a habit of becoming disconnected from its turbulent origins; this was observed by Sheila Whiteley in the 1970s in her study of progressive rock and the counterculture. She remarks about Pink Floyd: 'Political and social confrontation had become fragmented; subjective experience had degenerated into play power, which had little purchase other than an irreverent and often irrelevant questioning of authority, materialism and capitalism.'[2] We are witnessing this yet again with dance culture (Thornton, 1995).

Additionally, the 'shelf life' of the average pop artist is notoriously short; it has this in common with that other great route out of the ghetto of listlessness and poverty, sport. For a female artist in Britain (although not so much in the US), the shelf life is even shorter. Whether in the world of pop, which by definition is disposable, or rock, which by definition is male:

> *Rock is a pedestal sport, as is being a Monarch – wherever possible a boy inherits the throne – females are not thought to be the stuff worship/idols are made for/of. Girls are expected to grovel in the mezzanine while the stud struts his stuff up there, while a girl with the audacity to go on stage is jeered, sneered and leered up to. Rock and roll is very missionary, very religious, very repressive. A guitar in the hands of a man boasts 'cock' – the same instrument in female hands (to a warped mind) screams 'castration'.*[3]

Women in pop and rock are isolated and objectified, decontextualised from the social relations often believed to be the woman's forte. 'In order to be 'successful' in a male-dominated society, they must package themselves (or be packaged, as in advertising images) as objects amenable to control by men.' [4] So against this background, how possible is it to present oneself as relevant to the contemporary woman?

I found it hard to believe that all the energy in evidence during the period spanned by punk (roughly 1976 to 1982) simply dissipated. For that reason I have embarked on research to find out exactly what happened to halt the process of change that would have seen our pop music charts overflowing with female instrumentalists and performers voicing the concerns of our community, whether personal or apocryphal in nature. For that reason too, I decided to embark on this performance/installation project, as a continuation of my songwriting 'life-in-progress', choosing seven stages of life (some past, some future) to explore in song, interspersed with films contributed by women artists. An Arts Council bursary had funded training in computer music programming, which meant I could write fully orchestrated arrangements for the material. Having gained the luxury of access to public expression in the vernacular of pop, why should I not continue to articulate responses to life in the pop idiom, and to articulate 'nowness'[5] from the female perspective?

The pop generation is constantly being defined by high-profile observers who claim objectivity; it is rarely defined by its participants. Women have often been written out of pop history, by not being written in. Our subcultural value has been low, swept aside in the maleness of countercultural activity; this is ironic, given that one of the main elements of popular music of all types is the 'outward expression' of emotions. This was not even noticed until Angela McRobbie (1980) patiently unravelled male texts on youth and style,[6] pointing out the patriarchal nature of theorists on youth culture, and the absence of the subcultural history of our gender.

This is not to deny the symbiotic relationship pop has with its audience. Changes in pop aesthetics traditionally emanate from two main areas – from a subcultural 'hot-spot' in a community, however loosely defined, and/or advances in technology (which also entail a symbiotic relationship between manufacturers and users). As a performer, I want to communicate with not only 'people like me', but also a new audience. It has been important to articulate, through pop lyrics, feelings that I cannot communicate solely through the written word, to an audience who may not enjoy the written word as much as they do pop music. In pop, the audience are expected to interpret the shorthand of lyrics according to their own experience, by adding the singer's interpretation to the literal meaning of the words:

> *In songs, words are the sign of a voice ... Songs are more like plays than poems; song words work as speech and speech acts, bearing meaning not just semantically, but also as structures of sound that are direct signs of emotion and marks of character. Singers use non-verbal as well as verbal devices to make their points ... It is not just what they sing, but the way they sing it, that determines what singers mean to us and how we are placed, as an audience, in relationship to them.*[7]

Thus simple lyrics can convey meaning to an audience equipped by common personal experience to translate them. There is inherent risk in the apparent simplicity of pop lyrics: as Greil Marcus notes, 'Pop's anti-intellectualism, its rejection of a cultural heritage in favour of instant creativity, means that its executants rely entirely on their instincts ... given such inspiration, it can move brilliantly and rapidly along a path but eventually it finds itself facing a brick wall without the means to climb over it'.[8] Marcus' use of the word reflects the

Jukebox Jury hit-and-miss element of pop as well as its potential lack of engagement with 'the big world out there'; and innovation, essential to the fashion aspect of pop, is inhibited by the necessity of communication to the 'Common People'.[9] With this song cycle I approach another conundrum too; Angela McRobbie's observation that 'it's not so much that girls do too much too young; rather, they have the opportunity to do too little too late'[10] applies to many of us whose hedonistic youth convinced us that we had achieved parity before our time had come. We believed in revolution through our participation in music; with hindsight, we now understand our 'progress' from a different perspective:

> *Feminists know that if rock/pop was really revolutionary, they would be embraced as the greatest rebels of all-real rebels, the genuine article, not just another piece in the jigsaw of popular ephemera ... When they're fourteen, girl fans attract a lot of study and analysis ... But what happens when we grow up and become a minority in the audience for 'serious music'?*[11]

Because protest songs have become rather a joke in the music world – the image of the girly singer with guitar and billowing dress is never far from our minds, I chose to compose in a genre normally associated with glamour (big band swing) and to perform in neo-drag, a mutton dressed as a ram, instead of a mutton dressed as a lamb. This could arguably be seen as an attempt to 'elevate' the persona to another plane, a denial of gender.[12] But according to Virginia Woolf, we have two choices as women, either to 'demur to men' (I am only a woman) or to protest that we are 'as good as a man'.[13] Maybe it is an excuse to fall back on Marcia Citron's assertion about the female classical composer: 'She actually stands as a double outsider to society. As one who does not fit the assumption of the male professional she bears an added layer of otherness ... this can cause serious problems of self-image and identity',[14] and to carry this further into the realms of the composer/performer.

The songs are therefore rooted in a puzzlement with what pop music is actually for. Is it for revolution? Is it for sale? Is it for the perpetuation of youth culture? Is it purely an egotistical exercise? Is it useful? Is it for its creators, its audience or both? Is it subject to potential command by women?

Ultimately, I believe it is all these things, with the additional feature of 'fixing time'; it provides a sonic diary in the memory, and our varying engagement with different idioms of pop and rock music at different times of our lives has been one of the main experiential features of the common lifestyle during the twentieth century, and will remain so in the twenty-first. The lyrics, and the timbre of my voice, describe a forty year old woman's perspective, as I readjust and reassess the body I have inhabited since birth, and will inhabit till I die. There are other voices, too, in the songs, collected in the field, sampled, and mixed into the orchestration: sometimes the listener barely hears them, and sometimes they are featured prominently. Dialogue from the films will also be incorporated into the music in this way, to 'open up' the way the songs are interpreted. All pitched musical instruments mimic timbres of the human voice, here, instead of attempting to emulate instrumentation with the samples of the human voice, I will use different snapshots of speech to question the closed nature of what we create. John Shepherd (in Music as Social Text) discusses many issues that confront the female gender in the field of music, including our battle with male hegemony regarding post-Renaissance educated 'culture':

> It not only appears transparent, clear, lucid and self-evident, but also, therefore, self-sufficient. It is a culture that one either takes or leaves. To use Walter Benjamin's terminology, it does not absorb the spectator, it can only be absorbed by the spectator. In Barthes' terms it is 'readerly' (Barthes 1975). It does not allow the spectator in to complete the meaning.[15]

William Upton had used the concept of 'absorption' in 1880 when explaining the inability of women to compose music, 'Man controls his emotions, and can give an outward expression of them. In woman they are the dominating element, and so long as they are dominant she absorbs music'.[16] It is my intention to show that as females we

213

can be active in the creation of music and film, and also essential that the audience for this performance/installation does not merely 'absorb' what they see and hear, that they complete the meaning for themselves. This is why the music is only a part of the story. Voxpop Puella is also a collection of ideas voiced by seven female artists working with film or video, each exploring one of the seven ages of woman, in which I take the role of the live anchorwoman, linking their films with seven songs. The brief for each film-maker was simple, a short three or four minute movie (the duration chosen as the average length of a pop single), dealing with a given 'age'. There were no group meetings, and no editorial interference from myself, beyond managing the project so that the artists did not overlap as far as the 'age' chosen. Each contributor's film reflects their own 'voice', some of them telling us of emotional battles, some of them closed texts, some playing with the idea of truth. In most, we hear a woman's voice, either as a voice-over or speaking directly to us: in Eggs, Fish and Blood, the scratch animation by Joan Ashworth, there is no voice; we hear the human heartbeat and wonder whether this is a film about a mother, or about a child. Akiko Hada shadows an Icelandic teenager getting ready to go out clubbing; Jane Prophet arouses our curiosity with a confession that we find puzzling, playing with our curiosity. Gina Birch, framed by a glamorous red velvet couch, talks warmly and humorously about being forty; Gail Pearce invites us to her fiftieth birthday party, sparkling and singing. Finally, we meet Winifred, great grandmother of film-maker Charlotte Worthington, as spirited as an eight-year-old despite her 102 years.

We 'change focus' with the personality of each subject and each director, sometimes confused by the changes in our engagement with what we are seeing and hearing. The film-makers have been chosen for their strength of artistic vision. I did not want Video Diaries a la BBC 2. I believe that we see in these films a stage in the erosion of performative gender, Judith Butler describes gender as 'an identity tenuously constituted in time – an identity instituted through a stylised repetition of acts'.[17] Each artist plays with/to the camera, as if to challenge Berger's assertion that: 'To be born a woman is to be born, within an allotted and confined space, into the keeping of men ... A woman must continually watch herself. She is almost

continually accompanied by her own image of herself. Whilst she is walking across the room or whilst she is weeping at the death of her father, she can scarcely avoid envisaging herself walking or weeping'.[18]
The women we see represented in these films are funny, independent, articulate; where sexuality appears it is uncontrived and unexpected. The optimism is surprising, and provides an unanticipated thread throughout.

Many of us work with commercial constraints on our imagination, and part of the exercise for all of us was to say as much as possible in a simple medium in a tiny gap of time. Feedback from the artists during the filming gave the impression that there would be marked similarities of style, but this was not the case, indeed, the results were fascinating in their difference of approach. In 1979 I had been involved in publishing fanzines, the editorial policy had been 'hands-off', and consequently the resulting fanzines were a patchy conglomeration of styles and quality. I had wanted this project to have the same origi- nating ethos but here have had confidence that the visual images would be of a higher quality with an inherent defiance, which would be part of their appeal.

In 1988 I saw a performance at the Edinburgh International Festi- val that had more impact on me than anything I've seen before or since. The French director Jerome Deschamps had assembled a group of retired music hall stars who had been languishing, forgotten, in an old people's home. He formed a company and developed his idea into a performance entitled 'Les Petits Pas', using the performances of the elderly people, and employing actors to make a comedy of the indifference of their companions. The result was breathtaking – as an audience we were reminded of the fact that our right to be seen and heard, whether performers or not, extends for the whole of our lives. As Deschamps remarked: 'There is something worthy of con- sideration in the gestures that elderly people make, gestures informed by the whole of their past, which disclose tiny pieces of their lives. Through these gestures it is possible to glean tiny fragments of their lives, and I find that extraordinary'.[19]

The audience was pulled from the comedy of the actors to the ten- derness and agility displayed in the performing skills of the retired artists, always aware of the boundaries between artifice and reality:

215

these were not people 'acting old' – they were real people, performing their 'acts' as they had done many years ago. It is this tension between empathy and knowingness, and the idea of access to a voice, that I hope to continue developing in Voxpop Puella.

I hope that, by writing this, I have not blunted the impact of the performance, or replaced an experience with words. I want to perform the show in venues that stretch my capabilities as a performer while reaching people who can see themselves somewhere in the piece. I am approaching Arts Centres primarily, but hope to take the completed performance into a Retirement Home too. Ultimately, there will be different formats from the low-tech living-room version to the hi-tech, high-specification Arts venue version, with a projected backdrop of abstract human cells while the songs are being performed. Will people find the material stimulating? The ideal critical response to Voxpop Puella would be a counter-performance of some sort, to avoid a contemporary paradox, 'What seems to happen now is that in any article there is the moment of closure, the signifying process is seized up, there is a lurch into meaning – 'logocentrism' – and a succeeding article will develop from it. Or a text simply calls out for a supplementary text; there is no border'.[20] Perhaps it is naive to court an empirical reaction from an audience, but ultimately I believe that Voxpop Puella fulfils Sontag's definition:

> *A work of art encountered as a work of art is an experience, not a statement or the answer to a question. Art is not only about something, it is something. A work of art is a thing in the world, not just a text or a commentary on the world ... the knowledge we gain through art is an experience of the form or style of knowing something rather than a knowledge of something in itself.*[21]

APPENDIX

Film-makers in alphabetical order:

Joan Ashworth – conception/birth: 'Eggs, fish and blood'

Gina Birch – 40 years

Rachel Davies – titles

Darlajane Gilroy (proposed) – 70 years

Akiko Hada – teenager

Gail Pearce – 50 years: 'So this is 50'

Jane Prophet – 10 years

Charlotte Worthington – 102 years: '102'

Contributors

DR. TESSA ADAMS is responsible for the MA in Psychoanalytic Studies and the MA Psychotherapy and Society in the Unit of Psychotherapeutic Studies at Goldsmiths College University of London. Her PhD in the History and Theory of Art, awarded in 1993, focused on the 'Creative Experience and the Authenticity of Psychoanalytical Discourse'. Her recent publications include chapters in *The Sublime in Psychoanalysis: Archetypal Psychology and Psychotherapy*, ed., P.Clarkson, (Whurr 1997) and *The Creative Feminine: a Study of Kristeva*, eds., C. Heenan and B. Seu, (Sage 1998) Recently, Tessa has given a series of talks at the Hayward Gallery and the Serpentine Gallery.

STEVIE BEZENCENET is the course director for the MA in Design and Media Arts at the University of Westminster. She has published widely on issues of photographic history and criticism, but has recently turned to a photographic visual practice and is particularly interested in issues of gender / space relations. Stevie is a member of 'Cutting Edge'.

JOS BOYS works as an educational technologist at De Montfort University, UK. Her background is in architectural design, critical practice and theory. She is currently completing a PhD on the historical development of ideas about architectural space and is interested in exploring the inter-relationships between conceptual, material and virtual worlds. Jos is a member of 'Cutting Edge'.

PHILIPPA GOODALL was the guest editor for *Digital Desires*. For five years since 1993 she was Director of Photography Programming at Watershed Media Centre in Bristol. During this time she curated the exhibitions programme, initiated conferences and new media projects, oversaw the content and administration of Watershed's short course and gallery education programme and had management responsibility for the Photography and New Media Department. From April 1999 she combined freelance and consultancy work with a part-time role at Watershed carrying out project based new media development. Philippa was a member of 'Cutting Edge'.

LUCIA GROSSBERGER-MORALES was born in Catavi, Bolivia, in 1952. Her family emigrated to the United States when she was three and a half because of political turmoil. Her work deals with cross-cultural issues and experience of being an immigrant and a woman. She uses the computer to reach out and involve others in a creative collaborative experience. She has designed light shows and kaleidoscopic installations for rock concerts, Lallapalooza, Peter Gabrial's WOMAD (World of Music and Dance), and the US Festival sponsored by Steve Wozniak.

She graduated from the University of Southern California with a degree in anthropology in 1974 and a MS in Instructional Technology. Her work has been seen in computer shows such as SIGGRAPH, ISEA and SIGCHI Interactive Media Festivals, as well as in museums, galleries and film shows around the country.

SIAN HAMLETT graduated from The University of Westminster with a BA Honours degree in Contemporary Media Practice. She then went on to gain an MA in Psychoanalytic Studies from Brunel University. Sian is a freelance film and multimedia Producer/Director working on a diverse portfolio of high level innovative productions.

MAREN HARTMANN received an MA in Media Studies both from the University of Sussex and from the Free University in Berlin. Currently employed as a Teaching Assistant at the University of Westminster in the School of Communication, Design and Media, she is working on her PhD about the Cyberflâneur and the use of metaphors to describe online experiences. She worked as a Research Officer for a European Research Network (European Media, Technology and Everyday Life) at the University of Sussex.

PENNY HARVEY works in the Department of Social Anthropology, University of Manchester and is concerned with ethnographic and theoretical work on the politics of communication. Her main interest is in how identities are constituted through concrete practices of objectification, such as language use, visualisation and technological innovation. Author of *Hybrids of Modernity: Anthropology, the Nation State and the Universal Exhibition* (Routledge 1996) and editor of *Sex and Violence: Issues in Representation and Experience* (Routledge 1994).

JACKIE HATFIELD is an artist using film, digital image, sound and performance. She makes expanded moving image installation and single screen work. She is currently Lecturer in Comtemporary Media Practice at the University of Westminster where she is completing a practice-based PhD in authorship, audience and artwork, developing installations with moving images, random access delivery systems and multimedia authoring. She lectures in video, theory and practice at the University of Westminster. Jackie is a member of 'Cutting Edge'.

SARAH KEMBER is a lecturer at Goldsmiths College in New Technologies of Communication. Her book *Virtual Anxiety* was published by Manchester University Press in 1998. Sarah is a member of Cutting Edge.

SANDRA KEMP is Professor of English Literature and Cultural Theory at the University of Westminster. She writes on fiction, film, feminist theory and the performing arts. Her publications include Kipling's *Hidden Narratives* (1987), *Italian Feminist Thought* (1989), *Edwardian Fiction: An Oxford Companion* (1996) and *Feminisms: A Reader* (1997). Her recent publications include work on the archive as hypertext and on the relationship between the self, the face and art. She is currently working on a book on the face. Sandra is a member of Cutting Edge.

ERICA MATLOW worked for many years as a practising graphic designer before moving into education. She has an MA from the Institute of Education in the Sociology of Education and is currently researching, for her PhD, the impact that new technology has had on graphic design education. She is a Principal Lecturer at the University of Westminster, and coordinator of the Graphic Information Design BA (Hons). Erica is coordinator for the Research Group 'Cutting Edge' and co-editor of *Desire by Design* and *Digital Desires*. She is the Co-Vice Chair of CADE (Computers in Art and Design Education).

ANGELA MEDHURST Since graduating from the University of Westminster in 1995 Angela Medhurst has worked extensively in the new media industry as a freelance programmer and project manager. Projects have included web sites for BT, the BBC, Microsoft, Virgin Music as well as arts projects for Film & Video Umbrella and Camerwork. Anj is a researcher in digital practice and theory at the University of Westminster and is also involved in doctoral research there. Anj was also a co-founder of 'technowhores', a practice/theory collective involved in various conferences and exhibitions since 1995.

SHERRY MILLNER Internationally exhibited at such sites as the World-wide Video Festival, the Museum of Modern Art, the Whitney Biennial, in major festivals at Dallas, Barcelona, London, Tokyo, Glasgow, Amsterdam, Montreal, and Atlanta, among many others, Sherry Millner's award-winning films and videotapes have been widely praised. Her installations and photomontage projects have been represented in many books. Associate Editor of the media journal 'Jump Cut', Millner has taught at California Institute of the Arts, Rutgers University, Antioch College, and Hampshire College. Currently she is Chair of the Performing and Creative Arts Dept. at Staten Island, CUNY.

DR GABY PORTER was responsible for developing, interpreting and caring for Manchester's Museum of Science and Industry collections and contrib-

221

uted to the management and growth of the Museum through leading the development of new galleries and, most recently, ICT strategy and projects. She is an advocate of accessible, relevant, multi-cultural and multi-sensory approaches to exhibition-making, and of the use of imaginative and 'scenographic' techniques in museums. She developed a feminist analysis of museums in her doctoral thesis, a study of gender and representation in history museums in Great Britain (1994).

DR. JANE PROPHET is a British artist working with video and digital media. After graduating from Sheffield Hallam University in 1987, where she studied Fine Art, she went on to complete an MA in Electronic Graphics at Coventry University. Her interest in digital systems and artists' use of computer imaging became the focus of a PhD at Warwick University. Jane is a lecturer in Fine Art at the Slade School of Art and a researcher at the University of Westminster. She uses new media technologies in the production of Internet sites, CDROMs and large scale interactive installations. The award winning web-site 'TechnoSphere' reflects her interest in landscape and artificial life. Jane is a member of 'Cutting Edge'.

HELEN REDDINGTON is a songwriter who also writes music for film, TV and video. She started as a bass-player in a punk band, eventually recording many sessions for 'The John Peel Show' and other Radio One shows. Her innovative band 'Helen and the Horns' signed up with RCA records in the 1980s. In 1993 she completed an MA in Performance Art at Middlesex University, where she developed the performance 'Hermes', a song-cycle based on the seven deadly sins, which was later performed at the Tristan Bates Theatre in London's West End as well as community venues such as Jackson's Lane. She now lectures in music production and runs the thesis module on the BA Hons Commercial Music course at the University of Westminster, and continues to write music for film and TV. Most recently she has advised the National Year of Literacy on Lyrics. Helen is an associate member of 'Cutting Edge".

REFERENCES

Section 1
INNERTEXTUALITIES
CYBERBODIES AND PSYCHOANALYSIS

Sandra Kemp
Technologies of the face
1. John Liggett, *The Human Face,* London: Constable, 1974, p.275. See also John Brophy, *The Human Face Considered,* London: George Harrap and Co. Ltd., 1962.
2. See Joanna Woodall, *Portraiture: Facing the Subject,* Manchester and New York: Manchester University Press, 1991.
3. Alan Sekula. 'The Body and the Archive', *October,* 39 (Winter 1986).
4. Michael Ondaatje, *The English Patient,* London: Picador, 1993.
5. Charles Baudelaire, *Art and Artists,* transl. E.Keller, London: Macmillan, 1987.
6. See Richard Brilliant, *Portraiture,* London: Reaktion, 1991; Melissa E.Feldman, *Face-Off: The Portrait in Recent Art,* Philadelphia: Institute of Contemporary Art, 1996; Katherine Hoffmann, *Concepts of Identity: Historical and Contemporary Images of Self and Family,* New York: Harper Collins, 1996; Marcia Pointon, *Portraiture and Social Formation in Eighteenth Century England,* New Haven and London: Yale University Press, 1977.
7. Milan Kundera, *Immortality,* New York: Grove Weidenfeld, 1991.
8. Quoted in Mary Panzer, 'A Glow of Success', *The Independent on Sunday,* 25 October 1998.
9. Quoted in *The Times,* 3 January 1998.
10. John Hull, *Touching the Rock: An Experience of Blindness,* New York: Random House, 1992.
11. Jonathan Cole, *About Face,* Massachusetts: Massachusetts Institute of Technology, 1998.
12. 'Imaginaria', ICA, June 1998.
13. Quoted in Anna Murphy, 'Nobody's Perfect', *The Observer,* 7 June 1998.
14. 'The Nazis', The Photographer's Gallery, London WC2, August-September 1998.
15. Quoted in Anna Murphy, op. cit., p.18. My italics.
16. *The Big Issue,* 15-21 February 1999, p.25. My italics.
17. Quoted in *The [Guardian] Editor*, 30 January 1999, p.13; taken from the Jan./ Feb. issue of *Psychology Today.*
18. George Eliot, *The Mill on the Floss,* 3 vols. (1860: London: Chatto and Windus, 1986).
19. Simon Hoggart's Diary, *The Saturday Review, The Guardian,* 16 January 1999
20. *Village Voice Literary Supplement,* October 1991, p.22.
21. 'Fa(e)ces of the World': http:/www.bryne.dircon.co.uk/faeces/index/htm.
22. For further information on the FPRG and bibliographies, see University of Westminster web site.
23. Vicky Bruce and Andy Young, *In the Eye of the Beholder: The Science of Face Perception.* Oxford: Oxford University Press, 1998.
24. Ibid. pp.142.
25. See *New Scientist,* 27 February 1999, No. 2715, p.43.
26. There is a vast literature on physiognomics, but see: E.C.Evans, 'Physiognomics in the Ancient World', *Transactions of the American*

Philosophical Society, 59 (1969); Graeme Tytler, *Physiognomy in the European Novel,* Princeton: Princeton University Press, 1982); E.H.Gombrich, 'The Mask and the Face: The Perception of Physiognomic Likeness in Life and Art', in M. Mandelbaum (ed.) *Art, Perception and Reality,* Baltimore: Sparrow, 1972; Maud Gleason, *Making Men,* Princeton: Princeton University Press, 1995; Mary Cowling, *The Artist as Anthropologist: The Representation of Type and Character in Victorian Fiction,* Cambridge, Cambridge University Press, 1989.

27. J.K.Lavater, *Essays on Physiognomy: for the promotion of the love and knowledge of mankind,* London: Robinson, 1793.
28. S.J.Gould, *The Mismeasure of Man,* Harmondsworth, Penguin, 1989.
29. Quoted in ibid.
30. For polygenists, see J.B.Davis's *Thesaurus Craniorum* a catalogue of 1,500 skulls; Morton's collection of skulls which ranks races by the average size of their brains; e.g. American Indians in *Crania Americana,* 1839; skulls from Egyptian tombs, *Crania Egyptica,* 1844).
31. E.H.Gombrich, 'The Mask and the Face: The Perception of Physiognomic Likeness in Life and Art' in M. Mandelbaum, (ed.), *Art, Perception and Reality,* Baltimore: Sparrow, 1995.
32. Mary Cowling, *The Artist as Antropologist: The Representation of Type and Character in Victorian Fiction,* Cambridge: Cambridge University Press, 1989.
33. See also Mina Loy's 'Auto-Facial Construction', 1919, reprinted in Roger L.Conover, (ed.) *The Lost Lunar Baedeker,* London: Carcanet, 1997.
34. Quoted in Mary Cowling, op.cit.
35. Alan Sekula, 'The Body and the Archive', *October,* 39 (Winter 1986).
36. See *The Times,* 8 March 1999.
37. See *The Daily Mirror,* 19 September 1997, *The Guardian,* 19 September 1997.
38. *Time* Magazine, October 16, 1995.
39. Claudia Johnson, 'Fair Maid of Kent? The arguments for and against the Rice Portrait of Jane Austen', *TLS* 13 March 1998, pp.12-19.
40. 'Electonically Yours', The Museum of Photography, Tokyo, Summer 1998.
41. See also the work of Bruce Neuman and Chuck Close.
42. *Aesthetic Plastic Surgery,* 17:99-102. 1993 (Springer Verlag, New York Inc).
43. Cherry Norton and Richard Woods, 'Stars in their Eyes', *The Sunday Times,* 25 October 1998.
44. Quoted in *New Scientist,* 27 February 1999 No.2715, p.43.
45. Charles Bell, *The Anatomy and Philosophy of Expression,* 3rd.edn. London: George Bell, 1844.
46. Ludwig Wittgenstien, *Remarks on the Philosophy of Psychology,* Chicago: Chicago University Press, 1980.
47. Maurice Merleau Ponty, *The Primacy of Perception,* Evanston, IL: Northwestern University Press, 1964.
48. Emmanuel Levinas, *Collected Philosophical Papers ,* Dordrecht, Netherlands: Kluer.
49. Sigmund Freud, 'The Uncanny' (1919), *Standard Edition of the Complete Psychological Works of Sigmund Freud,* (ed.) James Strachey, vol.17, London: Hogarth Press, 1955.
50. Jonathan Miller, *On Reflection,* London: National Gallery Publications, 1998
51. Sarah Kofman, *The Enigma of Woman: Woman in Freud's Writings,* Ithaca: Cornell University Press, 1985. See Elizabeth Bronfen, *Over Her Dead Body: Death, Femininity and the Aesthetic* Manchester and New York: Manchester University Press, 1992.
52. Jonathan Cole, *About Face,* Massachusetts: University Of Massachusetts Press, 1988.

53. The best example of this is the Gothic novel with its coded references to sexology, evolutionary models etc.
54. E.O.Wilson, *Consilience: The Unity of Knowledge,* Little, Brown, 1998.
55. Quote in *The Independent,* 15 June 1998, p.24.
56. Richard Neave and John Pragg, *Making Faces,* London: British Museum Press, 1998.
57. Quoted in *The Guardian,* 27 May 1998, p.9.
58. *New Scientist,* my italics.
59. Quoted in Charles Jonscher, 'We're Only Human ...', *The Observer Review,* 24 January, 1999.
60. Charles Jonscher, *Wiredlife: Who are We in the Digital Age?* London: Bantam Press, 1998.
61. Shakespeare, *Macbeth.*
62. Maurice Grosser, *The Painter's Eye,* New York: Mentor, 1951.
63. Richard Brilliant, *Portraiture,* London: Reaktion Books, 1991. See also Tom Standage, 'Why the quality of wisdom is non-digital', *Financial Times,* 20 February 1999, IV.
64. Jenny Diski, *The Dream Mistress,* Guernsey, Channel Islands: Phoenix, 1996, pp.13; 16–17.

Jane Prophet and Sian Hamlett
Sordid Sites: The internal organs of a cyborg

1. The distinction between clinical structures and overt behaviour is crucial to the Lacanian approach. Just because someone engages in voyeuristic behaviour does not necessarily mean that they have a perverse structure. See Dylan Evans, *An Introductory Dictionary of Lacanian Psychoanalysis,* London, USA and Canada: Routledge, 1996.
2. Jacques Lacan, 'Subversion de sujet et dialectique de desir dans l'inconscient freudien', in Jacques Lacan, *Ecrits,* Paris: Seuil, 1960 ('The subversion of the subject and the dialectic of desire in the Freudian unconscious'), trans. Alan Sheridan, in Jacques Lacan, *Ecrits: A Selection,* London: Tavistock, 1977.
3. Jacques Lacan, *Le Seminaire. Livre XI. Les quatre concepts fondamentaux de la psychanalyse, 1964,* (ed.) Jacques-Alain Miller, Paris: Seuil, 1973 [The Seminar. Book XI. The Four Fundamental Concepts of Psychoanalysis, trans. Alan Sheridan, London: Hogarth Press and The Institute of Psychoanalysis, 1977.
4. Jean Clavreul, The Perverse Couple, trans. Stuart Schneiderman, in Stuart Schneiderman (ed.), *Returning to Freud: Clinical Psychoanalysis in the School of Lacan,* New Haven and London: Yale University Press, 1967.
5. Jacques Lacan, *Le Seminaire. Livre XI. Les quatre concepts fondamentaux de la psychanalyse,1964,* (ed.) Jacques-Alain Miller, Paris: Seuil, 1973 [The Seminar. Book XI. The Four Fundamental Concepts of Psychoanalysis, trans. Alan Sheridan, London: Hogarth Press and The Institute of Psychoanalysis, 1977.
6. Jacques Lacan, *Le Seminaire. Livre VII. L'ethique de la psychanalyse, 1959 – 60,* (ed.) Jacques-Alain Miller, Paris: Seuil, 1986 [The Seminar. Book VII. The Ethics of Psychoanalysis, 1959-60 trans. Dennis Potter, with notes by Dennis Potter, London: Routledge, 1992].
7. Jacques Lacan, 'Le Stade du miroir comme formateur de la fonction du Je', in Jacques Lacan, *Ecrits,* Paris: Seuil, 1966 ['The mirror stage as formative of the function of the I'], trans. Alan Sheridan, in Jacques Lacan, *Ecrits: A Selection,* London: Tavistock, 1977.
8. Lacan explores the notion of the fragmented body and its relationship to the mirror stage in 'L'agressivite en psychanalyse', in Jacques Lacan, *Ecrits,* Paris: Seuil, 1966 ['Aggressivity in psychoanalysis', trans. Alan Sheridan, in Jacques Lacan, *Ecrits: A Selection,* London: Tavistock, 1977.

9. See 7.

10. Jacques Lacan, *Le Seminaire. Livre III. Les psychoses, 1955-56,* (ed.) Jacques-Alain Miller, Paris: Seuil, 1981 [The Seminar. Book III. The Psychoses, 1955 – 56, trans. Russell Grigg, with notes by Russell Grigg, London: Routledge, 1993].

Sarah Kember
Get ALife: Cyberfeminism and the politics of artificial life

1. ECAL '97, CyberLife Technology, The Arts Council of England, Brighton Media Centre (1997) *LikeLIFE,* (exhibition catalogue).

2. Stuart Millar, Frankenstein's successor: a purple spider with a mind of its own that likes to go walkabout, *The Guardian,* Monday July 28, 1997.

3. ECAL '97, CyberLife Technology, The Arts Council of England, Brighton Media Centre (1997) *LikeLIFE* (exhibition catalogue).

4. Ibid.

5. CyberLife Creatures, CyberLife Technology Limited, Cambridge, 1997.

6. Margaret Boden, (ed), *The Philosophy of Artificial Life,* Oxford: Oxford University Press, 1996.

7. Ibid.

8. Ibid.

9. Ibid.

10. Sadie Plant, 'The Virtual Complexity of Culture' in George Robertson, Melinda Mash, Lisa Tickner, Jon Bird, Barry Curtis and Tim Putnam (eds.) *FutureNatural*, London: Routledge, 1996.

11. Boden, Margaret (ed.) (1996) *The Philosophy of Artificial Life*, op.cit.

12. Ibid.

13. Thomas S. Ray, 'An Approach to the Synthesis of Life' in Margaret Boden (ed.) *The Philosophy of Artificial Life*, Oxford: Oxford University Press, 1996.

14. Rosi Braidotti, *Nomadic Subjects*, New York, Columbia University Press, 1994, Sarah Franklin, 'Post-Modern Procreation. Representing Reproductive Practice', *Science as Culture*, 3,4,17., 1993. Sarah Kember, *Virtual Anxiety. Photography, New Technologies and Subjectivity*, Manchester: Manchester University Press, 1998.

15. Sherry Turkle, '*Life on the Screen. Identity in the Age of the Internet*, London: Phoenix, 1997.

16. Pattie Maes,'Artificial Life Meets Entertainment: Lifelike Autonomous Agents' in Lynn Hershman Leeson (ed.) *Clicking In. Hot Links to a Digital Culture,* Seattle: Bay Press, 1996.

17. Sherry Turkle, '*Life on the Screen. Identity in the Age of the Internet*, London: Phoenix, 1997.

18. Pattie Maes,'Artificial Life Meets Entertainment: Lifelike Autonomous Agents' in Lynn Hershman Leeson (ed.) *Clicking In. Hot Links to a Digital Culture,* Seattle: Bay Press, 1996.

19. Ibid.

20. Ibid.

21. Ibid.

22. Ibid.

23. Ibid.

24. Ibid.

25. Alison Adam, *Artificial Knowing. Gender and the Thinking Machine*, London: Routledge, 1998.

26. Ibid.

27. Ibid.

28. Donna J. Haraway, *Modest_Witness@Second_Millenium. FemaleMan©_ Meets_OncoMouse™*, London: Routledge, 1997.

29. Ibid.
30. Sadie Plant, 'The Future Looms: Weaving, Women and Cybernetics', in Mike Featherstone and Roger Burrows (eds.) *Cyberspace, Cyberbodies, Cyberpunk*, London: Sage, 1995.
31. Ibid.
32. Ibid.
33. Ibid.
34. Ibid.
35. Ibid.
36. Faith Wilding, and the Critical Art Ensemble 'Notes on the Political Condition of Cyberfeminism', URL: http://www.desk.nl/nettime).
37. VNS Matrix 'All New Gen', in Mathew Fuller (ed.) *Unnatural. Techno-theory for a contaminated culture*, London: Underground, 1994.
38. Sarah Kember, 'Feminist Figuration and the Question of Origin', in George Robertson, Melinda Mash, Lisa Tickner, Jon Bird, Barry Curtis and Tim Putnam (eds.) *FutureNatural*, London: Routledge, 1996.
39. Donna J. Haraway, *Modest_Witness@Second_Millenium. FemaleMan©_ Meets_OncoMouse™*, London: Routledge, 1997.
40. Ibid.
41. Ibid.
42. Ibid.
43. Ibid.
44. Ibid.
45. Ibid.
46. Ibid.
47. Ibid.
48. Ibid.
49. Ibid.
50. Ibid.
51. Ibid.
52. Ibid.
53. Ibid.
54. Ibid.
55. Ibid.
56. Ibid.
57. Allucquere Rosanne Stone, *The War of Desire and Technology at the Close of the Mechanical Age*, Cambridge, Massachusetts and London: England MIT Press, 1996., Sarah Kember, *Virtual Anxiety. Photography, New Technologies and Subjectivity*, Manchester: Manchester University Press, 1998.
58. Ibid.
59. Ibid.
60. Rosi Braidotti, 'Cyberfeminism with a Difference', *New Formations. Technoscience*, 29, 1996.
61. Stefan Helmreich, 'Replicating Reproduction in Artificial Life: Or, the Essence of Life in the Age of Virtual Electronic Reproduction' in Sarah Franklin and Helena Ragone (eds.), *Reproducing Reproduction. Kinship, Power and Technological Innovation,* Philadelphia: University of Pennsylvania Press, 1998.
62. Ibid.
63. Ibid.

Tessa Adams
Whose Reality is it Anyway? A Psychoanalytic Perspective

1. Sherry Turkle, *Life on the Screen,* London, Weidenfeld and Nicolson, 1996.
2. Melanie Klein proposes an early phase of infancy in which paranoid phantasy mitigates the fear of annihilation, termed the 'Paranoid-Schizoid position'. An established psychoanalytic view of adult psychosis maintains that severe psychological disturbances are a manifestation of this 'primitive' early paranoid-schizoid functioning. Melanie Klein, *Love, Guilt and Reparation and Other Works 1921-1945,* London: Virago Press, 1988.
3. Here 'phantasy' is distinguished from 'fantasy', since it represents the psychoanalytic understanding of the unconscious dynamic of the 'inner world' of relationship and desire. Fantasy, as more commonly used, in this context would refer solely to conscious desire.
4. MI5 is the code branch name of the British Ministry of Intelligence which is renowned for its capacity to seek out treason.
5. Mary Jacobus, Evelyn Fox Keller, Sally Shuttleworth (eds.), *Body/Politics: Women and the Discourses of Science,* London and New York: Routledge, 1990
6. Corporeal autonomy in this case refers to the belief that we 'own' our bodies, deriving from the humanistic assumption that the body is the 'temple' of the spirit. This is a view that promotes the concept of self-realisation in which the integration of mind and body result in our becoming 'whole'.
7. Sarah Kember, *Virtual Anxiety: photography, new technologies and subjectivity,* Manchester and New York: Manchester University Press, 1998.
8. Ibid.
9. This is an area of technological advancement that has attracted much feminist debate, see *Camera Obscura: No. 28 January 1992 No. 29 May 1992.*
10. Carol Stabile, 'Shooting the Mother: Fetal Photography and the Politics of Gender', *Camera Obscura: No. 28, January 1992 No. 29 May 1992.*
11. Ibid.
12. Sherry Turkle, *Life on the Screen,* London: Weidenfeld and Nicolson, 1996
13. John Harris, *Clones, Genes and Immortality.*
14. R.C. Lewotin, *The Doctrine of DNA: Biology as Ideology,* London: Penguin Books, 1992.
15. Marion Milner's view is that the infant in the very early months develops the illusion that not only are its body products 'love gifts' that it gives to its mother, but that it created her/himself and the world around her/him. Marion Milner, *Suppressed Madness of Sane Men: Forty Years of Exploring Psychoanalysis,* London: Routledge, 1990.
16. Orlan is a French artist whose artwork consists of the orchestration, documentation and video presentation of a series of surgical operations that she has designed to alter her face. She develops the 'blue print' for the operations by computer generated collated images which the surgeons are asked to follow. During the operations she interacts with the surgeon, medical crew and the process of surgery, using a monitor to follow the procedures.
17. 'Meat' is a derogatory Cyberpunk term used to describe the body. To be 'meat' is human and inferior to the technological supremacy of the Cyborg. See Claudia Springer 'The Pleasure of Interface': *Screen* 32.3 Autumn 1991.
18. The debates on the status of the Virtual Body are discussed extensively by Springer (Ibid.) and Simon Penny, 'Virtual Reality and the Completion of the Enlightenment Project' in *Culture on the Brink ,* Gretchen Bender, Timothy Duckrey, (eds.) Seattle: Bay Press, 1994.
19. The reference here is to the project ALife that aims to evolve biotechnological 'creatures' which replicate the processes of living organisms. See Sarah Kember's chapter in this book.

20. Ibid.
21. Richard Dawkins, *The Blind Watchmaker,* New York: W. W. Norton and Co., 1986.
22. Sherry Turkle, 1996 *Life on the Screen,* London: Weidenfeld and Nicolson, 1996.
23. Ibid.
24. The 'Symbolic' in Lacanian terms is a modality of significance associated with law and rationality, and includes verbal communication – sign, syntax, and grammatical constraint. Jacques Lacan, *Ecrits: A Selection,* London: Tavistock Publications, 1977.
25. Sherry Turkle, 1996 *Life on Screen,* London: Weidenfield and Nicolson, 1996.
26. 'Potential Space' is a Winnicottian term for the quality of maternal environment essential for the infant's capacity to be creative, without which autonomy suffers. Donald Winnicott, *Playing and Reality,* London: Pelican, 1971.
27. 'The (la) Semiotic' is a Kristevian term for the realm of the Maternal which includes pre-verbal bodily communication, the infant echolias or vocalising prior to sign or syntax. It is her view that this repressed realm of maternal 'signification' erupts through language (the patriarchal order) exemplified in poetic discourse. Julia Kristeva, *Desire in Language: A Semiotic Approach to Literature and Art,* Oxford: Blackwell, 1980.
28. Acting out is a psychiatric term to denote behaviour that does not subscribe to social codes. It is usually disruptive, aggressive and dismissive of concern for others.
29. Of Jung's Four Archetypes the 'Trickster-Figure' is mercurial, manipulative, cheating and seductive, embraced by the group while creating havoc. Carl Gustave Jung, *Four Archetypes,* London: Routledge, 1972.
30. Andrew Ross, *Strange Weather,* London: Verso, 1991.
31. Ibid.

Section 2
OUTERTEXTUALITIES
INTERNAL SITES AND EXTERNAL EXPERIENCES

Sherry Millner
Wired for Violence

1. *Scenes from the Micro-War* is distributed by Video Data Bank, 112 S. Michigan Av., Chicago IL 60603.
2. Dave Grossman, *On Killing,* New York: Little, Brown and Company, 1995.

Angela Medhurst
Shop 'Til You(r Connection) Drop(s): Considering the electronic supermarket

1. Royal Town Planning Institute; *Planning for shopping into the 21st century: the report of the Retail Planning Working Party;* London; Royal Town Planning Institute, 1988.
2. See Juliet Webster, *'Shaping Women's Work – Gender, Employment and Information Technology',* London: Longman, 1996 for a thorough introduction to the gender and employment politics of new technologies in the workplace. Also Eileen Green, Jenny Owen; Den Pain, eds., *Gendered by Design?: Information technology and office systems,* London: Taylor & Francis; 1993.
3. See Jennifer Terry, Melodie Calvert (eds.), *Processed Lives, Gender & Technology in Everyday Life',* Ch. 5, London: Routledge, 1997 and various works by Allucquère Rosanne Stone, Sadie Plant et al.

4. See Anne Balsamo, *Technologies of the Gendered Body,* Durham: Duke University Press, 1996.
5. See Judy Wajcman, *'Feminism confronts technology'*, Cambridge, Polity, 1991.
6. Rachel Bowlby, *'Supermarket Futures'* in Pasi Falk, Colin Campbell, *The shopping experience,* London: Sage, 1997.
7. Peter K. Lunt, Sonia M. Livingstone, *Mass consumption and personal identity: everyday economic experience,* Buckingham: Open University Press, 1992.
8. Daniel Miller, 'Could Shopping Ever Really Matter?', in Pasi Falk, Colin Campbell, *The shopping experience,* London: Sage, 1997.
9. Peter K. Lunt, Sonia M. Livingstone; 1992.
10. See (among many others) Julien Stallabrass, *Gargantua manufactured mass culture*, London; Verso, 1996.
11. Rob Shields, *Lifestyle shopping: the subject of consumption*, London: Routledge, 1992.
12. Ibid.
13. Ibid.
14. Dympna McGill, *New Media Age,* 'E-credit? That'll do nicely sir ...', 25/6/98.
15. Peter K. Lunt, Sonia M. Livingstone; 1992.
16. This paper does not allow for the inclusion of detailed usage statistics or access demographics. See http://www.redsquare.co.uk/survey/index.htm for a regularly updated internet user survey. Week 18/2/99: 'Typical Profile: A single male aged 22 – 30 who owns his own home and car, lives in or near a metropolitan area and may have an income exceeding 40k pa. If he reads a paper, it's likely to be *The Guardian* – he's well educated (to degree level) and holds a senior position in the banking, finance or business services sectors. He's been using the net for a year or more, and uses it every day at work for Web browsing and email, spending more than 5 hours each week online'.
17. Nigel Cope, *Retail in the Digital Age,* London: Bowerdean, 1996.
18. Pasi Falk, Colin Campbell, 'Introduction', *The shopping experience,* London: Sage. 1997.
19. Nigel Cope, *Retail in the Digital Age,* London: Bowerdean, 1996.
20. Rachel Bowlby; 'Supermarket Futures'; in Pasi Falk, Colin Campbell, *The shopping experience,* London: Sage, 1997.
21. Nigel Cope, *Retail in the Digital Age,* London: Bowerdean, 1996.
22. Ibid.
23. Vicki August, 'Retail feels competitive crunch of e-commerce', *Information Week,* 22/7/98 – 4/8/98.

Maren Hartmann
The Cyberflâneuse: Metaphors and reality in virtual space

1. Iain Sinclair, *Lights Out For the Territory.* London: Granta Books, 1997.
2. Nick Melczarek, Sur les Champs-Webess: a Web flâneur/boulevardier at large, 1998 http://nersp.nerde.ufl.edu/~nickym/9010a.html (19/01/98).
3. Franz Hessel, *Ein Flâneur in Berlin,* Berlin: das Arsenal, 1984.
4. Mark Stefik, *Internet Dreams. Archetypes, Myths, and Metaphors.* Cambridge: Mass. & London, MIT Press, 1996.
5. Merriam-Webster, WWWebster Dictionary. http://www.m-w.com/cgi-bin/dictionary (17/8/98).
6. Priscilla Parkhurst Ferguson, *Paris As Revolution – Writing the 19th-Century City.* Berkeley, Los Angeles & London: University of California Press, 1994b.

7. Rob Shields, Fancy footwork. Walter Benjamin's notes on *flânerie*. In Keith Tester, (ed.), *The Flâneur*. London/New York: Routledge, 1994.
8. Walter Benjamin, *Charles Baudelaire A Lyric Poet in the Era of High Capitalism*. London/New York: Verso, 1997.
9. Ibid.
10. William Mitchell, *City of Bits. Space, Place and the Infobahn,* Cambridge, Mass. & London: MIT Press, 1995.
11. Steven Goldate, The Cyberflâneur Spaces and Places on the Internet. http://vicnet.net.au/~claynet/flaneur.htm 1997.
12. Stephan Porombka, Der elektronische Flâneur als Kultfigur der Netzkultur. http://www.icf.de/softmoderne/1/Hyperkultur/text/Po_1_1.html 1997.
13. William Mitchell, *City of Bits. Space, Place and the Infobahn,* Cambridge, Mass. & London: MIT Press, 1995.
14. Bowlby, 1992; Buck-Morss, 1986; Ferguson, 1994; Nava, 1996; Pollock, 1988; Wilson, 1992; Wolff, 1990.
15. Janet Wolff, The Invisible Flâneuse: Women and the Literature of Modernity, in Janet Wolff, *Feminine Sentences: Essays on Women and Culture,* Cambridge/Oxford: Polity Press, 1990.
16. One of the most important publications describing the characters visible in the streets of the 19th century Paris – the 'physiologies' – mentions only the Grisettes, but no other female character (this is the collection of thirteen booklets (plus reference to three more) that are currently available in the British Library).
17. Janet Wolff, 'The Invisible Flâneuse: Women and the Literature of Modernity', in Janet Wolff, *Feminine Sentences: Essays on Women and Culture,* Cambridge/Oxford: Polity Press, 1990.
18. Deborah Epstein Nord, *Walking the Victorian Steets – Women, Representation, and the City.* Ithaca and London: Cornell University Press, 1995.
19. Mica Nava, 'Modernity's Disavowal: Women, the city and the department store', in Mica Nava, & Alan OShea, (eds.), *Modern Times. Reflections on a century of English modernity.* London & New York: Routledge, 1996.
20. Nua Internet Surveys. http://www.nua.ie/surveys/ 1999.
21. GVU's 10th WWW User Survey – General Demographics Survey – Gender. http://www.gvu.gatech.edu/user_su...8-04/reports/1998-04-General.html 1999.
22. Rena Tangens, Ist das Internet männlich? Über Androzentrismus im Netz. In Stefan Bollmann, & Christiane Heibach, *Kursbuch Internet*. Mannheim: Bollmann Verlag, 1996.
23. PENet, Prostitutes Education Network, http://www.bayswan.org/intro.html 1998
24. Julian Dibbell, A Rape in Cyberspace or How an Evil Clown, A Haitian Trickster Spirit, Two Wizards, and a Cast of Dozens Turned a Database Into a Society. http://ftp.lambda.moo.mud.org/pub/MOO/papers/VillageVoice.txt 1993.
25 Wilson, 1991, Wolff, 1990, Ferguson, 1994b, Nord, 1995.
26. Allucquère Rosanne Stone, *The War of Desire and Technology at the Close of the Mechanical Age.* Cambridge, Mass. & London: MIT Press, 1995.
27. McRae, 1997; Stivale, 1997; Turkle, 1995.
28. Kathy Rae Huffmann, & Margarethe Jahrmann, Mailing lists as tools to build electronic communities. http://www.heise.de/tp/english/pop/event_2/4118/1.html 1998.
29. Bettina Lehmann, Internet – (r)eine Männersache? Oder: Warum Frauen das Internet entdecken sollen. In: Bollmann, Stefan & Heibach, Christiane: *Kursbuch Internet,* Mannheim: Bollmann Verlag, 1996.

30. Benjamin drew a parallel between prostitution and play – in both you face fate (Benjamin, 1989:612). And Bauman (in Tester, 1994:138-157) describes the flâneur as the homo ludens. It is also interesting to note the use of arcades today as the place for computer games, mainly frequented by teenage boys.
31. Nua Internet Surveys. http://www.nua.ie/surveys/ 1999.
32. Bettina Lehmann, Internet – (r)eine Männersache? Oder: Warum Frauen das Internet entdecken sollen. In: Bollmann, Stefan & Heibach, Christiane: *Kursbuch Internet,* Mannheim: Bollmann Verlag, 1996.
33. Susan Buck-Morss,The Flâneur, the Sandwichman and the Whore: The Politics of Loitering. In: *New German Critique,* No.39. Fall 1986.
34. Walter Benjamin, *Charles Baudelaire: A Lyric Poet in the Era of High Capitalism.* London / New York: Verso, 1997.
35. Kathy Rae Huffmann, & Margarethe Jahrmann, Mailing lists as tools to build electronic communities. http://www.heise.de/tp/english/pop/event_2/4118/1.html 1998.
36. Griselda Pollock, *Vision and Difference. Femininity, feminism and histories of art,* London & New York: Routledge, 1988.
37. Lucy Kimbell, Cyberspace and subjectivity: the MOO as the home of the postmodern flâneur. http://www.soda.co.uk/Flaneur/flaneur1.htm 1997.
38. Mica Nava, Modernity's Disavowal Women, the city and the department store. In: Nava, Mica & Alan OShea, (eds.), *Modern Times. Reflections on a century of English modernity.* London & New York: Routledge, 1996.
39. Priscilla Parkhurst Ferguson, 'The *flâneur* on and off the streets of Paris', in Keith Tester, *The Flâneur,* London & New York: Routledge, 1994a.
40. Ibid.
41. Nua Internet Surveys. http://www.nua.ie/surveys/ 1999.
42. GVU's 10th WWW User Survey – General Demographics Survey – Gender. http://www.gvu.gatech.edu/user_su...8-04/reports/1998-04-General.html 1999.
43. Nua Internet Surveys. http://www.nua.ie/surveys/ 1999.

Penny Harvey and Gaby Porter
Infocities: From information to conversation

1. Penny Harvey was doing ethnographic research on the Infocities team, as part of a wider ESRC funded project, 'social contexts of virtual Manchester'. This project also involved fellow anthropologist Sarah Green and historian John Agar and was carried out as part of the ESRC's 'Virtual Society?' programme. Gaby Porter was a member of the Infocities team, as Curatorial Services Manager in the Museum of Science and Industry in Manchester.
2. Interviewees names have been changed. The ages given are approximate.
3. This is the usage adopted by Terry and Calvert (1997). Jennifer Terry and Melodie Calvert, (eds.), *Processed Lives: Gender and Technology in Everyday Life,* London: Routledge, 1997.
4. See the excellent book by Francesca Bray (1997) which takes this view of technology to examine gender and textile production in China. Francesca Bray, *Technology and Gender: Fabrics of Power in Late Imperial China,* Berkeley: University of California Press, 1997.
5. People who have advanced this argument include Plant (1997) – Sadie Plant, *Zeros and Ones: Digital Women and The New Technologies,* London: Fourth Estate, 1997 – and Rothschild (1983). J. Rothschild, (ed), *Machina Ex Dea: Feminist Perspectives on Technology,* New York: Pergamon, 1983.
6. See for example Webster (1996:25), "I view women's relationship to technology as one of exclusion through embedded historical practice, reinforced and reproduced in contemporary work settings". J. Webster, *Shaping Women's Work: Gender, Employment and Information Technology,* Harlow: Longman, 1996.

7. See Rowbotham (1995). Sheila Rowbotham, Feminist approaches to technology: women's values or a gender lens? in Swasti Mitter and Sheila Rowbotham, (eds.), *Women Encounter Technology: Changing Patterns of Employment in the Third World,* London: Routledge, 1995.

8. Cynthia Cockburn, and Susan Ormrod, *Gender and Technology in the Making,* London: Sage, 1993.

9. Ibid.

10. See for example McNeil (1987) – Maureen McNeil, ed, *Gender and Expertise,* London: Free Association Books, 1987 – and Gill & Grint (1995). Rosalind Gill, and Keith Grint, (eds.), *The Gender-Technology Relation: Contemporary Theory and Research,* London: Taylor & Francis, 1995.

11. Donna Haraway, *Simians, Cyborgs and Women: The Reinvention of Nature,* New York: Free Association Books, 1991.

12. Cynthia Cockburn, and Susan Ormrod, *Gender and Technology in the Making,* London: Sage, 1993.

13. See Cockburn on domestic technologies (1983, 1985). Cynthia Cockburn, *Brothers: Male Dominance and Technological Change,* London, Pluto Press. 1983, Cynthia Cockburn, *Machinery of Dominance: Women, Men and Technical Know-how,* London: Pluto Press, 1985.

14. A point also made by Webster (1996), J. Webster, *Shaping Women's Work: Gender, Employment and Information Technology,* Harlow: Longman. 1996.

15. Keith Grint and Steve Woolgar, 'On some failures of nerve in constructivist and feminist analyses of technology', in Rosalind Gill and Keith Grint, (eds.), *The Gender-Technology Relation: Contemporary Theory and Research,* London: Taylor & Francis, 1995.

16. Cynthia Cockburn, and Susan Ormrod, *Gender and Technology in the Making,* London: Sage, 1993.

Jos Boys
Windows on the World:
Architecture, identities and new technologies

1. The South Bank in London 'happened' to be the space I was occupying when I first began working on this piece and seemed a good location for testing its ideas through personal observation. I would like to imply that the selection was thus relatively random. The truth is, I use the South Bank a lot in the summer as an outdoor living room when I'm in London. This, of course, affects the ways I articulate its qualities.

2. These technologies were selected on the basis of a single photographic journey taken along the South Bank from Waterloo Bridge to the OXO Tower one weekday afternoon in September 1998.

3. Donna J. Haraway, *Modest_Witness@ Second Millenium. FemaleMan© Meets_Oncomouse™,* London and New York: Routledge, 1997.

4. See Sarah Kember, this volume.

5. The term 'commonsense' is used here not as a straightforward set of agreed beliefs but as a bundling of assumptions based on ideas we think with, rather than think *about*. See Errol Lawrence, 'Just plain commonsense: the "roots of racism"', Centre for Contemporary Cultural Studies *The Empire strikes back. Race and Racism in 70s Britain,* London: Hutchinson, 1982.

6. Jos Boys, *'Concrete Visions? Architectural knowledge and the production and consumption of buildings'.* University of Reading unpublished PhD, forthcoming.

7. Peter Reyner Banham, *The New Brutalism: Ethic or Aesthetic?* Architectural Press, 1966.

8. John Elsom and Nicholas Tomalin, *The History of the National Theatre,* Jonathan Cape, 1978.
9. Jos Boys 'Women and public space' in *Matrix, Making Space: Women and the Man-Made Environment,* London: Pluto Press 1984 and 'From Alcatraz to the OK Corral: Post-war housing design' in Judy Attfield and Pat Kirkham (eds.), *A View from the Interior,* London: Women's Press (2nd edition) 1991.
10. Rosalyn Deutsche, *Evictions: Art and Spatial Politics,* New York: MIP Press, 1996.
11. See competition entries in *Designing the Future of the South Bank,* London: Academy Editions, 1994.
12. For example, Jonathan Hill, (ed.), *Occupying Architecture,* London and New York: Routledge, 1998, Duncan McCorquodale et al., (eds.), *Desiring Practices,* London: Black Dog Publishing, 1996., Diane Ghirardo, (ed.), *Out of Site,* Seattle: Bay Press, 1991 and Thomas A. Dutton and Lian Hurst Mann (eds.), *Reconstructing Architecture,* Minneapolis: University of Minnesota Press, 1996.
13. Iain Borden, 'Body Architecture : Skateboarding and the creation of super-architectural space' in Hill, (ed.), *Occupying Architecture,* op cit.
14. Henri Lefebvre, *The Production of Space,* Oxford, Blackwell, 1991.
15. Borden op cit.
16. This information is from conversations with an art student who is also a skate boarder and who attended the second Cutting Edge conference at which an early version of this paper was presented.
17. Donna J. Haraway, *Modest_Witness@ Second Millenium, FemaleMan© Meets_ Oncomouse™,* London and New York: Routledge, 1997.
18. Pierre Bourdieu, *Distinction: a social critque of the judgement of taste,* London: Routledge 1986.
19. The CCCS spawned important cultural studies work on, for example, advertising (Judith Williamson, *Consuming Passions,* London: Marion Boyars 1986), product design (Dick Hebdige, *Hiding in the Light*), youth cultures, (Angela McRobbie) and the black diaspora (Paul Gilroy, *There ain't no black in the Union Jack).*
20. See, for example, Gillian Rose, *Feminism and Geography,* Cambridge: Polity Press 1993, Judith Butler, *Bodies that Matter*, New York and London: Routledge 1993 and Luce Irigaray, *Thinking the Difference,* London: The Athlone Press 1994.
21. Rosalyn Deutsche op cit.
22. Elizabeth Gorsz 'Bodies-Cities' in Beatriz Colomina, (ed.), *Sexuality and Space* Princeton: Princeton Architectural Press, 1992.
23. Jos Boys 'Positions in the Landscape' in Cutting Edge, (eds.), *Desire by Design: Bodies Territories and New Technologies,* London and New York: I. B. Tauris, 1999.
24. Philip Tabor 'Striking home: the telematic assault on identity', in Jonathan Hill ed., *Occupying Architecture,* London and New York: Routledge, 1998.
25. Ibid.
26. FAT (Fashion, Architecture, Taste) 'Contaminating Contemplation' in Jonathan Hill (ed.), *Occupying Architecture*, London and New York: Routledge, 1998.
27. For examples of such binary mapping see David Harvey, *The Condition of Postmodernity,* Cambridge, Massachusetts: Basil Blackwell, 1989.
28. For a more popularist version of the mobile phone as either threat or pleasure see David Newham 'Speak to me please' *Guardian Weekend* June 12, 1999.

234

29. Personal observation of mobile phones in use at the South Bank, September 1998
30. See, for example Rosalyn Deutsche, *Evictions: Art and Spatial Politics,* New York: MIT Press, 1996; Doreen Massey, *Space, Place and Gender,* Cambridge: Polity Press, 1994, Nancy Duncan (ed.), *Bodyspace,* London and New York: Routledge, 1996. Rosa Ainley (ed.), *New Frontiers of Space, Bodies and Gender,* London and New York: Routledge, 1998.
31. Doreen Massey, *Spatial Divisions of Labour,* Basingstoke: MacMillan, 1984.
32. Daniel Miller, *Material Culture and Mass Consumption,* Oxford: Blackwell, 1987 p215.
33. Walter Benjamin 'The work of art in the age of mechanical reproduction', quoted in Jonathan Hill (ed.), *Occupying Architecture,* London and New York: Routledge, 1998.

Section 3
INTERTEXTUALITIES
LANGUAGE, IDENTITY AND NEW TECHNOLOGIES

Erica Matlow
Women, computers and a sense of self

1. Langdon Winner quoted by Cynthis Cockburn, 'The Circuit of Technology' in Roger Silverstone and Eric Hirsch, (eds.), *Consuming Technologies,* Routledge: London, 1992.
2. Sherry Turkle, *Life on the Screen,* New York: Simon and Schuster, 1995.
3. Ibid.
4. 'Word processing is computer-assisted document preparation, which includes writing, editing, storing, proofing, printing and electronically transmitting documents written in one of the natural languages.' Michael Heim, *Electric Language,* New Haven and London: Yale University Press, 1987.
5. Sherry Turkle, *The Second Self,* New York: Simon and Schuster, 1984.
6. Cynthia Cockburn, and Susan Ormrod, *Gender and Technology in the Making,* London: Sage Publications Ltd, 1993.
7. Dale Spender, *Nattering on the Net,* Australia: Spinifex Press, 1995.
8. Sherry Turkle, 'Computers and the Human Spirit', in Ruth Finnegan, Graeme Salaman and Kenneth Thompson, (eds.), *Information Technology: Social Issues,* London: Hodder and Soughton, 1987.
9. 'The Macintosh interface, its simulated desktop, introduced a way of thinking that put a premium on surface manipulation and working in ignorance of the underlying mechanisms. The desktop's interactive objects, its anthropomorphised dialogue boxes in which the computer "spoke" to its user – these developments all pointed to a new kind of experience in which people do not so much command machines as enter into conversations with them.' Sherry Turkle, *Life on the Screen,* New York: Simon and Schuster, 1995.
10. Dale Spender, *Nattering on the Net,* Australia: Spinifex Press, 1995.
11. Ibid.
12. Ibid.
13. Richard Lanham, *The Electronic Word,* Chicago: The University of Chicago Press, 1993.
14. Sherry Turkle, *Life on the Screen,* New York: Simon and Schuster, 1995.
15. Ibid.
16. The Visible Language workshop at the Massachusetts Institute of Technology (MIT) was founded by Muriel Cooper and Ron MacNeil in 1975 to

investigate the inter-relationships between visual communication, design research and artificial intelligence.

17. Richard S.Wurman, *Information Architects*, Graphis Press Corp, Switzerland, 1996.
18. Richard Lanham, *The Electronic Word,* Chicago: The University of Chicago Press, 1993.
19. Dale Spender, *Nattering on the Net,* Australia: Spinifex Press, 1995.
20. Geoffrey Nunberg, *The Future of the Book,* University of California Press, Los Angeles, 1996.
21. Umberto Eco, 'Afterword' in Geoffrey Nunberg, *The Future of the Book,* University of California Press: Los Angeles,1996.
22. Sherry Turkle, 'Computers and the Human Spirit', in Ruth Finnegan, Graeme Salaman and Kenneth Thompson, (eds.), *Information Technology: Social Issues,* London: Hodder and Soughton, 1987.
23. Sherry Turkle, *Life on the Screen,* New York: Simon and Schuster, 1995.
24. Dale Spender, *Nattering on the Net,* Australia: Spinifex Press, 1995.
25. AIGA Journal of Graphic Design Vol. 13 No. 1 1995, American Institute of Graphic Arts, NY.
26. Dale Spender, *Nattering on the Net,* Australia: Spinifex Press, 1995.
27. Ibid.
28. Margaret Wylie 1995 'No Place for a Woman', Digital Media 4, 8, January, in Dale Spender, *Nattering on the Net,* Australia: Spinifex Press, 1995.
29. Wendy Falkner and Erik Arnold (eds.) *Smothered by Invention,* London and Sydney: Pluto Press, 1985.
30. Eileen Green, Jenny Owen and Den Pain, (eds.), *Gendered by Design,* London: Taylor and Francis Ltd., 1993.
31. Ibid.
32. Sherry Turkle, *Life on the Screen,* New York: Simon and Schuster, 1995.
33. Ibid.
34. Ibid.
35. Dale Spender, *Nattering on the Net,* Australia: Spinifex Press, 1995.
36. Cynthia Cockburn, 'The Circuit of Technology' in Roger Silverstone and Eric Hirsch, (eds.) *Consuming Technologies,* London: Routledge, 1992.
37. Ibid.

Lucia Grossberger–Morales
Sangre Boliviana: Using multimedia to tell personal stories
1. Eduardo Galeano, Memory of Fire: Genesis, New York: Pantheon Books, 1982.

Jackie Hatfield
Disappearing Digitally? Gender in the digital domain
1. The empirical statistical reality suggests that arguments of empowerment are overly optimistic, for example in 98/99 1.72% of the global population has access to the Internet : NUA Internet surveys: Nua Homepage http://www.nua.ie/.
2. Luce Irigary, *Je, Tu, Nous Toward a Culture of Difference*, London: Routledge, 1993.
3. By the mid eighteenth century 'women's ability to read and write trailed behind men's (by twenty to twenty-five percentage points in this era.', p 139, A History of Their Own, Women in Europe from Prehistory to the Present, Vol.II, Bonnie S. Anderson and Judith P. Zinsser, 1990, Penguin Books.
4. Sherry Turkle, *Life on the Screen. Identity in the Age of the Internet*, London: Phoenix 1997.

5. Ibid.
6. Ibid.
7. Denos C.Gazis The Evolving Resource in *The Future of Software*, Derek Leebaert, (ed.),The MIT Press, 1996; See also Paul E. Ceruzzi, *A Modern History of Computing*, The MIT Press, 1998.
8. Luce Irigaray, *Je, Tu, Nous Towards a Culture of Difference*, London: Routledge, 1993.
9. Ibid.
10. Dale Spender, *Man Made Language*, London: Pandora, 1992 ed.
11. Sadie Plant, *Zeros and Ones: digital women and the new technoculture*, London: Fourth Estate, 1997.
12. From a study based on data provided by from a telephone survey conducted from 1996 to January 1997. 'April 17 1998: A study by Vanderbilt University has revealed the extent of the disparity between whites and blacks with regard to both the Internet and PC ownership in the US. While 41 million whites have accessed the Internet only 5 million African Americans have gone online.' & 'The disparity was most obvious in households with incomes below the national average of $40,000.' NUA Internet Surveys. 'Washington (July 28 1998 10.46 EDT http://www.nandotimes.com) The 'digital divide' widened over the past three years, as more wealthy than poor Americans acquired computers and connected to the Internet [] . The study found that while nationwide the percentage of households with computers climbed to 36.6% from 24.1%, gains lagged for people living in inner city, poor or minority households; from Nando.net.
13. 'National Telecommunications and Information Administration' 'Falling Through the Net II: New Data on the Digital Divide' Data from The US Bureau of the Census, US Department of Commerce'.
14. Ibid.
15. Nicolas Negroponte, *Being Digital*, Coronet Books, 1995.
16. Ibid.
17. Sadie Plant, *Zeros and Ones: digital women and the new technoculture*, London: Fourth Estate, 1997.
18. Margaret Wertheim, *The Pearly Gates of Cyberspace, A History of Space from Dante to the Internet*, Virago, 1999.
19. An early form of book was called a 'codex' The Chinese developed a printing press in around AD 600, and the Europeans around 1444. (Johan Guttenberg printed the Gutenberg Bible in 1456, this is generally considered to be Europe's first printed book.)
20. For discussion of the notion of commodities as a patriarchal construct see Chapter 9, *This Sex Which is Not One* Luce Irigaray; Cornell University Press, 1985.
21. Douglas C. McMurthie, *The Book The Story of Printing and Bookmaking* 1976; first edition 1927.
22. Introduction, *Medieval Women Writers* (ed.) Katharina M.Wilson, Manchester University Press, 1984.
23. There were some women noted to have been involved as printers, in the fifteenth century. In the house of Estienne, in Paris, there was Guyonne Viart, and daughter in law and scholar Perrette; Charlotte Guillard (Paris); later in the 18th century in the US, Mary Goddard, Ann Smith Franklin, and Elizabeth and Ann Timothy. The fact that Dinah Nuthead, attributed as the first woman in the US to be in charge of a printing office in 1695 when she succeeded her husband William Nuthead, relinquished this succession because of her illiteracy, is particularly significant to my argument.

24. For further discussion around the dissemination of misogynistic pamphlets and texts see *A History of their Own, Women in Europe from Prehistory to the Present Vol II,* Bonnie S. Anderson and Judith P. Zinsser; 1990, Penguin Books.

25. Dale Spender, *Man Made Language*, London: Pandora, 1992 ed.

26. Ibid. 'Broadly speaking, semantics refers to the meanings available within the language, while syntax refers to the form (the sentence structure) in which those meanings are conveyed.'

27. See also Luce Irigaray the French Linguist and Philosopher who argues for mechanisms to disrupt male oriented structures and Western oriented phallocentrism. *This Sex Which is Not One* Luce Irigaray, Cornell University Press, 1985; *Thinking the Difference for a Peaceful Revolution*, Luce Irigaray, The Athlone Press, 1994.

28. Luce Irigary, *Je, Tu, Nous Toward a Culture of Difference,* London: Routledge, 1994.

29. Luce Irigaray, *Thinking the Difference for a Peaceful Revolution*, The Athlone Press, 1994.

30. Ibid.

31. Ibid.

32. A term for code breakers who were dealing with computational mathematics.

33. It was fortunate for Ada Lovelace that her mother, Annabella Milbanke, had an active interest in education, (see p. 22 Dorothy Stein *Ada, A Life and a Legacy*) and invested in tutors for her including her own mathematics teacher, William Frend. After meeting Charles Babbage at a party in 1833 Lovelace was invited to see his 'Difference Engine' and inspired by its mathematical principles started to obtain information from various sources to understand it, studying with Mary Somerville the mathematician. After her marriage, Ada Lovelace continued her study with Augustus De Morgan, Professor of Mathematics at the University of London, all the while cultivating a professional relationship with Charles Babbage. Whilst translating Luigi Federico Menabrea's paper for the British Journal 'Taylor's Scientific Memoirs' (Menabrea was an Italian military engineer later to become Italy's Prime Minister) which explained how the Difference Engine worked, Ada added her own explanation of its mechanism. For more information on Ada Lovelace see *Ada, A Life and a Legacy* by Dorothy Stein.

34. Ada Lovelace, Edith Clarke (1883-1959), Rosa Peter (1905-1977), Grace Murray Hopper (1906-1992), Alexandra Illmer Forsythe (1918-1980) and Kay McNulty Mauchly Antolnelli.

35. Without the impetus during the Second World War period to investigate and break codes through mathematical computational research the mechanical computer would not have been developed so early. The huge number of costly research hours that were needed to reach this technological point would have been prohibitive without the military financial input, or at least, the position we are in today in terms of computing would have come much later.

36. In 1944 as a Lieutenant in the US Navy, Grace Hopper was assigned to the Bureau of Ordinance Computation Project at Harvard University, where between 1937 and 1944 Howard Aiken was working on the development of the Mark 1 machine. Aiken had wanted to develop a mechanical computational machine to aid him on his PhD for which he required to compute vast numbers of calculations. Coinciding with World War Two, the Mark 1 research provided a computational tool for the US Navy to calculate the angle of aim of new naval guns and also to determine the possible geographical consequences of dropping an atomic bomb. Hopper worked on programming the Mark 1 computer.

38. See for example Valerie Walkerdine, *Counting Girls Out*, Falmer, 1998.

238

39. 'Software is the fastest-growing industry in the United States. While the U.S. economy expanded by 30 percent in the years 1984-1994, the software sector proliferated by 269 percent, becoming larger than all but five manufacturing industries' p.5, *'The Future of Software'* Derek Leebaert, (ed.), The MIT Press, 1996.

40. Sherry Turkle, *Life on the Screen. Identity in the Age of the Internet*, London: Phoenix, 1997.

41. Ibid.

42. NUA Internet Surveys http://www.nua.ie – see footnote 44.

43. 'Naturalware: Natural-Language and Human-Intelligence Capabilities' Gustave Essig; from *The Future of Software,* (ed.) Derek Leebaert, Massachusetts: MIT Press, 1996.

44. Anna Vignoles quoted from *'The Guardian'*; Saturday February 1999 'Shirking maths at school just doesn't add up in your paypacket'.

45. *Education in the UK Facts and Figures*, Donald Mackinnon, June Statham and Margaret Hales.

46. NUA Internet Surveys: email : web@nua.ie – from a study by Prof.Tracy Camp of the University of Alabama: 1983-84 females accounted for 37.1% of computer science graduates and in 1993-94 the figure dropped to 28.4%.

47. Dale Spender; *Man-Made Language*, London: Pandora, 1992 ed.

48. The significance of commercial oriented software's restrictions/limitations to liberty and freedom are not lost on the founder of the GNU Project Richard Stallman. There is extensive information on the Internet on the GNU Project. A search for GNU will bring up hundreds of sites, the GNU manifesto written in 1985 by Richard Stallman is clearly marked. GNU is a free software alternative to the UNIX operating system that dominates the market. The GNU Project promotes the use of free shared software that is written collectively to run on the GNU operating system. 'Extracting money from users of a program by restricting their use of it is destructive because the restrictions reduce the amount and the ways that the program can be used' from GNU Manifesto and 'Free Software' is a matter of liberty, not price. To understand the concept, you should think of "free speech", not "free beer" Richard Stallman quoted from 'What is Free Software?' – GNU Project on the Internet site.
    Stallman's argument is that software development is so powerful, so commercially driven that freedom of speech for the user is restricted. He argues that three levels of freedom are infringed by software commodification: 1. Incriptions on software determine that it is impossible to study how the programme works, or to change it (the coder/author and the coding being the commodity); 2. Licensing discourages copying of software, some software cannot be copied; 3. The user cannot customise or adapt the programme to their own needs. From the position of a programmer like Richard Stallman, commercial development and sale of software implies restriction, not liberation.

49. Sept 1997 'According to Nancy Deyo of Purple Moon a company specialising in developing computer games for girls, rather than being uninterested, girls are bored with the computer games currently on the market. It seems that the Barbie Fashion Designer CDRom, which was an unexpected success, stumbled on a previously untargeted yet potentially huge market. In an effort to get girls online and get them buying, companies like Purple Moon are concentrating on storylines which allow girls to explore new worlds and communicate with each other. Girls are not motivated by games with gratuitously violent narratives but demand more sophisticated storylines from computer games.' NUA Surveys http://www.nua.ie/surveys.

50. 'Women's College Co-Edition' p.1.http://www.academic.org/work.html.

Helen Reddington
Voxpop Puella

1. Dave Laing, *One Chord Wonders,* Milton Keynes and Philadelphia: OUP, 1985.
2. Sheila Whiteley, *The Space Between the Notes,* London: Routledge, 1992.
3. Sarah Thornton, *Club Cultures,* Cambridge: Polity, 1995.
4. Julie Burchill and Tony Parsons, *The Boy Looked at Johnny,* London: Pluto Press, 1978.
5. John Shepherd, *Music as Social Text,* Cambridge: Polity, 1991.
6. Dave Stewart (ex-Eurythmics) uses this word to describe the necessity of capturing 'the moment' in pop music.
7. Angela McRobbie, *Settling Accounts with Subcultures; a Feminist Critique in On Record,* (ed.), Simon Frith and Andrew Goodwin, London and New York: Routledge, 1990, (orig. published 1980).
8. Ibid.
9. Greil Marcus, *In the Fascist Bathroom,* London: Viking, 1993.
10. *Common People:* the title of a hit song by Indie chart band Pulp, in which lyricist Jarvis Cocker simultaneously derides and admires the concept of 'popular culture'.
11. Angela McRobbie, *Settling Accounts with Subcultures; a Feminist Critique in On Record,* (ed.), Simon Frith and Andrew Goodwin, London and New York: Routledge, 1990, (orig. published 1980).
12. Gina Rumsey and Hilary Little, 'Women and Pop: a Series of Lost Encounters', in *Zoot Suits and Secondhand Dresses,* Angela McRobbie, (ed.), Basingstoke: Macmillan, 1989.
13. Ibid.
14. Virginia Woolf, *A Room of One's Own*, London, Chatto and Windus, 1984, quoted in Citron.
15. Marcia Citron, *Gender and the Musical Canon,* Cambridge: Cambridge University Press, 1993.
16. John Shepherd, *Music as Social Text,* Cambridge: Polity, 1991.
17. George Upton, *Woman in Music,* Boston: Osgood, 1880.
18. Judith Butler, 'Performative Acts and Gender Constitution: an essay in Phenomenology and Feminist Theory', in *Theatre Journal,* Dec 1985, Vol 40.
19. John Berger, 1977, *Ways of Seeing,* London: BBC and Penguin, 1977.
20. Jerome Deschamps, programme notes to *'Les Petits Pas',* Royal Lyceum Theatre, Edinburgh International Festival, 1988.
21. Paul Oldfield, 'After Subversion: Pop Culture and Power', in *Zoot Suits and Secondhand Dresses,* Angela McRobbie, (ed.), Basingstoke: Macmillan, 1989.
22. Susan Sontag, *A Susan Sontag Reader,* London: Penguin, 1982.

Bibliography

Alison Adam, *Artificial Knowing. Gender and the Thinking Machine*, London: Routledge 1998.

Gaston Bachelard, *The Poetics of Space*, Boston: Beacon 1969.

Peter Reyner Banham, *The New Brutalism: Ethic or Aesthetic?* London: Architectural Press 1966.

Roland Barthes, 'The Death of the Author' in *Twentieth Century Literary Theory*. K. M. Newton (ed.) London: MacMillan 1988.

Ellen Bass and Laura Davis, *The Courage to Heal,* New York: Harper & Row 1988.

Charles Baudelaire, *Selected Writings on Art and Artists,* London 1972.

Walter Benjamin, *Charles Baudelaire – A Lyric Poet in the Era of High Capitalism,* London/New York: Verso 1997.

Gretchen Bender and Timothy Druckrey (eds.), *Culture on the Brink,* Seattle: Bay Press 1994.

John Berger *Ways of Seeing*, London BBC and Penguin 1977.

Margaret Boden (ed.) *The Philosophy of Artificial Life*, Oxford: Oxford University Press 1996.

Pierre Bourdieu, *Distinction: The Sociology of Taste* London: Routledge 1984.

Rachel Bowlby, *Still Crazy After All These Years – Women, Writing and Psychoanalysis*. London & New York: Routledge 1992.

Jos Boys 'Women and public space' in Matrix, *Making Space: Women and the Man-Made Environment*, London: Pluto Press 1984.

Jos Boys 'From Alcatraz to the OK Corral: Post-war housing design' in Judy Attfield and Pat Kirkham (eds.) *A View from the Interior*, London: Women's Press (2nd edition) 1991.

Rosi Braidotti, *Nomadic Subjects*, New York: Columbia University Press 1994.

Rosi Braidotti, 'Cyberfeminism with a Difference', *New Formations. Technoscience, 29* 1996.

Francesca Bray, *Technology and Gender: Fabrics of Power in Late Imperial China,* Berkeley: University of California Press 1997.

Vicky Bruce and Andy Young, *In the Eye of the Beholder. The Science of Face Perception*. Oxford: Oxford University Press 1998.

Susan Buck-Morss, *The Dialectics of Seeing: Walter Benjamin and the Arcades Project*. Cambridge, Mass. & London: MIT Press 1989.

Susan Buck-Morss, The Flâneur, the Sandwichman and the Whore: The Politics of Loitering. in *New German Critique*. No.39 Fall 1986.

Julie Burchill and Tony Parsons,*The Boy Looked at Johnny,* London: Pluto Press 1978.

Judith Butler, 'Performative Acts and Gender Constitution: an essay in Phenomenology and Feminist Theory', in *Theatre Journal*, Dec., Vol. 40 1985.

Marcia Citron, *Gender and the Musical Canon* Cambridge: Cambridge University Press 1993.

Hélène Cixous and Mireille Calle-Gruber *Hélène Cixous, Rootprints: memory and life writing*, Routledge, London 1997.

Hélène Cixous, *Stigmata: Escaping Texts*, London: Routledge 1998.

Harry Cleaver, 'The "Space" of Cyberspace: Body Politics, Frontiers and Enclosures' *http://www.eco.utexas.edu:80/Homepages/Faculty/Cleaver/index.html* 1996.

CyberLife 'Creatures', Cambridge: CyberLife Technology Limited 1997.

Cynthia Cockburn, *Brothers: Male Dominance and Technological Change,* London: Pluto Press 1983.

Cynthia Cockburn *Machinery of Dominance: Women, Men and Technical Know-how* London: Pluto Press 1985.

Cynthia Cockburn and Susan Ormrod, *Gender and Technology in the Making*, London: Sage 1993.

Cynthia Cockburn, 'The Circuit of Technology' in Roger Silverstone and Eric Hirsch (eds.), *Consuming Technologies,* London: Routledge 1992.

Jonathan Cole, *About Face,* Massachusetts: Massachusetts Institute of Technology, 1998.

Nigel Cope, *Retail in the Digital Age,* London: Bowerdean 1996.

Ruth Cowan Schwartz, *More work for mother : the ironies of household technology from the open hearth to the microwave,* London: Free Association 1983.

Mary Cowling, *The Artist as Anthropologist: The Representation of Type and Character in Victorian Fiction,* Cambridge: Cambridge University Press 1989.

Richard Dawkins, *The Blind Watchmaker*, London: Penguin 1986.

Julian Dibbell, *A Rape in Cyberspace or How an Evil Clown, A Haitian Trickster Spirit, Two Wizards, and a Cast of Dozens Turned a Database Into a Society.* http://ftp.lambda.moo.mud.org/pub/MOO/papers/VillageVoice.txt (20/10/97) 1993.

Designing the Future of the South Bank, London: Academy Editions 1994.

Rosalyn Deutsche, *Evictions: Art and Spatial Politics*, New York: MIP Press 1996.

Richard Doyle, *On Beyond Living. Rhetorical Transformations of the Life Sciences*, Stanford: California Stanford University Press 1997.

Nancy Duncan (ed.) *Bodyspace*, London & New York: Routledge 1996.

Thomas A. Dutton and Lian Hurst Mann (eds.) *Reconstructing Architecture*, Minneapolis: University of Minnesota Press 1996.

John Elsom and Nicholas Tomalin, *The History of the National Theatre,* London: Jonathan Cape 1978.

Joshua M. Epstein and Robert Axtell, *Growing Artificial Societies. Social Science from the Bottom Up*, Washington D.C., Brookings Institution Press and Cambridge, Massachusetts: The MIT Press 1996.

Rachel Bowlby, 'Supermarket Futures' in *The shopping experience,* Pasi Falk and Colin Campbell (eds.) London: Sage 1997.

'Fa(e)ces of the World': http:/www.bryne.dircon.co.uk/faeces/index/htm.

Pasi Falk and Colin Campbell, *The shopping experience*, London: Sage 1997.

Wendy Falkner and Erik Arnold (eds.), *Smothered by Invention*, London and Sydney: Pluto Press 1985.

Melissa E. Feldman, *Face-Off: The Portrait in Recent Art,* Philadelphia: Institute of Contemporary Art 1996.

Priscilla Parkhurst Ferguson, *Paris As Revolution - Writing the 19th-Century City.* Berkeley, Los Angeles & London: University of California Press 1994.

Priscilla Parkhurst Ferguson, 'The flâneur on and off the streets of Paris' in Keith Tester (ed.) *The Flâneur,* London & New York: Routledge 1994.

Janet Frame, *An Angel At My Table*, London: Paladin 1987.

Sigmund Freud, 'The Uncanny' (1919), *Standard Edition of the Complete Psychological Works of Sigmund Freud,* (ed.) James Strachey, vol.17, London: Hogarth Press, 1955.

Eduardo Galeano, *Memory of Fire: Genesis*, New York: Pantheon Books 1982.

Diane Ghirardo (ed.) *Out of Site,* Seattle: Bay Press 1991.

Rosalind Gill and Keith Grint (eds.) *The Gender-Technology Relation: Contemporary Theory and Research*, London: Taylor & Francis 1995.

Steven Goldate, *The Cyberflâneur Spaces and Places on the Internet.* http://vicnet.net.au/~claynet/flaneur.htm (07/07/97) 1997.

Eileen Green, Jenny Owen and Den Pain (eds.) *Gendered by Design*, London: Taylor and Francis 1993.

Marcus Greil, *In the Fascist Bathroom,* London: Viking 1993.

Keith Grint and Steve Woolgar, 'On some failures of nerve in constructivist and feminist analyses of technology' in Rosalind Gill and Keith Grint (eds.) *The Gender-Technology Relation: Contemporary Theory and Research,* London: Taylor & Francis 1995.

Dave Grossman, *On Killing,* New York, Little, Brown and Company 1995.

GVU, *10th WWW User Survey – General Demographics Survey – Gender.* http://www.gvu.gatech.edu/user_su...8-04/reports/1998-04-General.html (08/04/99) 1999.

Donna Haraway *Modest_Witness@Second_Millenium. Female_Man©_Meets_OncoMouse™,* London and New York: Routledge 1997.

Donna Haraway, *Simians, Cyborgs and Women: the Reinvention of Nature,* New York: Routledge 1991.

Sandra Harding, *Whose Science? Whose Knowledge? Thinking from Women's Lives,* Ithaca: Cornell University Press 1992.

Jill Harsin, *Policing Prostitution in Nineteenth Century Paris,* Princeton: Princeton University Press 1985.

Katherine N. Hayles, 'Simulated Nature and Natural Simulations: Rethinking the Relation Between the Beholder and the World' in William Cronon (ed.) *Uncommon Ground. Toward Reinventing Nature,* New York & London: W.W. Norton and Co 1995.

Stefan Helmreich, 'Replicating Reproduction in Artificial Life: Or, the Essence of Life in the Age of Virtual Electronic Reproduction' in Sarah Franklin and Helena Ragone (eds.) *Reproducing Reproduction. Kinship, Power and Technological Innovation,* Philadelphia, University of Pennsylvania Press 1998.

Michael Heim, *Electric Language,* New Haven and London, Yale University Press 1987.

Franz Hessel, *Ein Flâneur in Berlin.* Berlin: das Arsenal 1984.

Jonathan Hill (ed.) *Occupying Architecture,* London and New York: Routledge 1998

Kathy Rae Huffmann and Margarethe Jahrmann, *Mailing lists as tools to build electronic communities.* http://www.heise.de/tp/english/pop/event_2/4118/1.html (07/09/98) 1998.

Luce Irigaray, *Je,Tu,Nous Towards a Culture of Difference,* London: Routledge 1993.

Mary Jacobus, Evelyn Fox Keller, Sally Shuttleworth (eds.), *Body/Politics: Women and the Discourses of Science,* London and New York: Routledge 1990.

Barbara Johnson, *A World of Difference,* Baltimore: The Johns Hopkins University Press 1987.

Charles Jonscher, *Wiredlife: Who are We in the Digital Age?* London: Bantama Press 1998.

Carl Gustave Jung, *Four Archetypes,* London: Routledge 1972.

Sarah Kember, 'Feminist Figuration and the Question of Origin', in George Robertson, Melinda Mash, Lisa Tickner, Jon Bird, Barry Curtis and Tim Putnam (eds.) *FutureNatural,* London: Routledge 1996.

Sarah Kember, *Virtual Anxiety. Photography, New Technologies and Subjectivity,* Manchester: Manchester University Press 1998.

Lucy Kimbell, *Cyberspace and subjectivity: the MOO as the home of the postmodern flaneur.* http://www.soda.co.uk/Flaneur/flaneur1.htm (20/10/97) 1997.

Melanie Klein, *Love, Guilt and Reparation and Other Works 1921 – 1945,* London: Virago Press 1988.

Sarah Kofman, *The Enigma of Woman: Woman in Freud's Writings,* Ithaca: Cornell University Press 1985.

Julia Kristeva, *Desire in Language: A Semiotic Approach to Literature and Art,* (eds.) Léon S. Roudiez. Trans. Thomas Gora, Alice Jardine, & Léon S. Roudiez, Oxford: Basil Blackwell 1984.

Jacques Lacan, *Ecrits: A Selection,* London: Tavistock Publications 1977.

243

Jacques Lacan, 'The Mirror Stage'.*A Critical and Cultural Theory Reader* in (eds.) Antony Easthope & Kate McGowan, Sydney: Allen & Unwin 1993.

Dave Laing, *One Chord Wonders,* Milton Keynes and Philadelphia: OUP 1985.

Christopher G Langton, 'Artificial Life' in Margaret Boden (ed.) *The Philosophy of Artificial Life*, Oxford: Oxford University Press 1996.

Christopher G. Langton (ed.) *Artificial Life. An Overview*, Cambridge, Massachusetts and London: England MIT Press 1997.

Richard Lanham, *The Electronic Word*, Chicago: The University of Chicago Press 1993.

Bruno Latour, and Steve Woolgar, *Laboratory Life: The Social Construction of Scientific Facts*, London: Sage 1979.

Bruno Latour, *Science in Action: How to Follow Scientists and Engineers Through Society*, Cambridge: Harvard University Press 1987.

Darian Leader and Judy Groves, *Lacan For Beginners,* London: Icon Books 1995.

Bettina Lehmann, 'Internet - (r)eine Männersache? Oder: Warum Frauen das Internet entdecken sollen' in Stefan Bollmann and Christiane Heibach, *Kursbuch Internet*. Mannheim: Bollmann Verlag 1996.

R.C. Lewotin, *The Doctrine of DNA: Biology as Ideology,* London: Penguin Books 1992.

Peter K. Lunt and Sonia M. Livingstone, *Mass consumption and personal identity: everyday economic experience*, Buckingham: Open University Press 1992.

Pattie Maes, 'Artificial Life Meets Entertainment: Lifelike Autonomous Agents' in Lynn Hershman Leeson (ed.) *Clicking In. Hot Links to a Digital Culture*, Seattle: Bay Press 1996.

Doreen Massey, *Space, Place and Gender* Cambridge: Polity Press 1994.

Duncan McCorquodale et al (eds.) *Desiring Practices*, London: Black Dog Publishing 1996.

Dympna McGill, 'E-credit? That'll do nicely sir...', *New Media Age,* 25/6/98. 1998.

Maureen McNeil (ed), *Gender and Expertise*, London: Free Association Books, 1987.

Shannon McRae, 'Flesh Made Word: Sex, Text and the Virtual Body' in David Porter (ed) *Internet Culture*, New York & London: Routledge 1997.

Angela McRobbie (ed.) , *Zoot Suits and Secondhand Dresses,* Basingstoke: Macmillan 1989.

Angela McRobbie, (orig. published 1980), 'Settling Accounts with Subcultures; a Feminist Critique' in Simon Frith and Andrew Goodwin (eds.), *On Record*, London and New York: Routledge 1990.

Nick Melczarek, *Sur les Champs-Webess: a Web flâneur/boulevardier at large*. http://nersp.nerde.ufl.edu/~nickym/9010a.html (19/01/98) 1998.

Merriam-Webster, *WWWebster Dictionary*. http://www.m-w.com/cgi-bin/dictionary (17/8/98) 1998.

Daniel Miller, *Material Culture and Mass Consumption,* Oxford: Blackwell 1987.

Jonathan Miller, *On Reflection,* London: National Gallery Publications 1998.

Laura Miller, 'Women and Children First: Gender and the Settling of the Electronic Frontier' in James Brook and Iain A. Boal (eds.) *Resisting the Virtual Life. The Culture and Politics of Information*, San Fransisco: City Lights Books 1995.

Marion Milner, *Suppressed Madness of Sane Men: Forty Years of Exploring Psychoanalysis,* London: Routledge 1990.

William Mitchell, *City of Bits. Space, Place and the Infobahn*, Cambridge, Mass. & London: MIT Press 1995.

Swasti Mitter and Sheila Rowbotham (eds.) *Women Encounter Technology: Changing Patterns of Employment in the Third World*, London: Routledge 1995.

244

Toril Moi (ed.) *The Kristeva Reader*. Trans. Seán Hand & Léon S. Roudiez, Oxford: Basil Blackwell 1984.

Mica Nava, 'Modernity's Disavowal: Women, the city and the department store' in Mica Nava and Alan OShea (eds.) *Modern Times. Reflections on a century of English modernity*, London & New York: Routledge 1996.

Deborah Epstein Nord, *Walking the Victorian Streets – Women, Representation, and the City*, Ithaca and London, Cornell University Press 1995.

Nua, Internet Surveys. http://www.nua.ie/surveys/ (08/04/99) 1999.

Geoffrey Nunberg, *The Future of the Book*, Los Angeles: University of California Press 1996.

PENet: *Prostitutes Education Network*. http://www.bayswan.org/intro.html (13/09/98) 1998.

Sadie Plant, *Zeros and Ones: Digital Women and The New Technologies*, London: Fourth Estate 1997.

Sadie Plant, 'The Future Looms: Weaving, Women and Cybernetics', in Mike Featherstone and Roger Burrows (eds.) *Cyberspace, Cyberbodies, Cyberpunk*, London: Sage 1995.

Sadie Plant, 'The Virtual Complexity of Culture' in George Robertson, Melinda Mash, Lisa Tickner, Jon Bird, Barry Curtis and Tim Putnam (eds.) *FutureNatural*, London: Routledge 1996.

Griselda Pollock, *Vision and Difference: Femininity, feminism and histories of art*, London & New York: Routledge 1988.

Maurice Merleau Ponty, *The Primacy of Perception*, Evanston, IL: Northwestern University Press 1964.

Stephan Porombka, *Der elektronische Flâneur als Kultfigur der Netzkultur*. http://www.icf.de/softmoderne/1/Hyperkultur/text/Po_1_1.html (24/06/97) 1997.

Thomas S. Ray, 'An Approach to the Synthesis of Life' in Margaret Boden (ed.) *The Philosophy of Artificial Life*, Oxford: Oxford University Press 1996.

Stuart Rock (ed.) *Electronic commerce: a real business guide*, London: Caspian 1997.

Joan Rothschild (ed.), *Machina Ex Dea: Feminist Perspectives on Technology*, New York: Pergamon 1983.

Royal Town Planning Institute, *Planning for shopping into the 21st century: the report of the Retail Planning Working Party*, London: Royal Town Planning Institute 1988.

Gina Rumsey and Hilary Little, 'Women and Pop: a Series of Lost Encounters' in Angela McRobbie (ed.) *Zoot Suits and Secondhand Dresses*, Basingstoke: Macmillan 1989.

Graeme Salaman and Kenneth Thompson, *Information Technology: Social Issues*, London: Hodder and Stoughton 1987.

Alan Sekula, 'The Body and the Archive', *October*, 39 (Winter 1986).

Susan Sellers (ed.) *The Hélène Cixous Reader*, London: Routledge 1997.

Sabine Schülting, (forthcoming), *Poisoning the Blood of the Nation: Viktorianische Verhandlungen urbaner Prostitution*. Würzburg: Königshausen & Neumann.

John Shepherd, *Music as Social Text*, Cambridge: Polity 1991.

Rob Shields; *Lifestyle shopping: the subject of consumption*, London: Routledge 1992.

Iain Sinclair, *Lights Out For the Territory*, London: Granta Books 1997.

Susan Sontag, *A Susan Sontag Reader*, London: Penguin 1982.

Dale Spender, *Nattering on the Net*, Australia: Spinifex Press 1995.

Mark Stefik, *Internet Dreams. Archetypes, Myths, and Metaphors*, Cambridge, Mass. & London: MIT Press 1996.

Charles J. Stivale, 'Spam: Heteroglossia and Harassment in Cyberspace' in David Porter (ed.) *Internet Culture*, New York & London: Routledge 1997.

Allucquere Rosanne Stone, *The War of Desire and Technology at the Close of the Mechanical Age*, Cambridge, Massachusetts & London, MIT Press 1996.

Rena Tangens, 'Ist das Internet männlich? Über Androzentrismus im Netz' in Stefan Bollmann and Christiane Heibach, *Kursbuch Internet*, Mannheim: Bollmann Verlag 1996.

Jennifer Terry and Melodie Calvert (eds), *Processed Lives: Gender and Technology in Everyday Life*, London: Routledge 1997.

Sarah Thornton, *Club Cultures*, Cambridge: Polity Press 1995.

Sherry Turkle, *The Second Self*, New York: Simon and Schuster 1984.

Sherry Turkle, 'Computers and the Human Spirit' in Ruth Finnegan, Graeme Salaman and Kenneth Thompson (eds.) *Information Technology: Social Issues*, London: Hodder and Stoughton 1987.

Sherry Turkle, *Life on the Screen, Identity in the Age of the Internet*, New York: Simon and Schuster 1995.

Sherry Turkle, *Life on the Screen. Identity in the Age of the Internet*, London: Phoenix 1997.

George Upton, *Woman in Music*, Osgood: Boston 1880.

VNS Matrix 'All New Gen', in Mathew Fuller (ed.) *Unnatural. Techno-theory for a contaminated culture*, London: Underground 1994.

Juliet Webster, *Shaping Women's Work: Gender, Employment and Information Technology*, Harlow: Longman 1996.

Sheila Whiteley, *The Space Between the Notes*, London, Routledge 1992.

Faith Wilding and the Critical Art Ensemble, *Notes on the Political Condition of Cyberfeminism*, URL:http://www.desk.nl/~nettime

E.O.Wilson, *Consilience: The Unity of Knowledge*, Little, Brown 1998.

Elizabeth Wilson, 'The Invisible Flâneur' in *New Left Review*. No.191. January/February 1992.

Elizabeth Wilson, *The Sphinx in the City. Urban Life, the Control of Disorder, and Women*, London: Virago Press 1991.

Donald Winnicott, *Playing and Reality*, London: Pelican 1971.

Ludwig Wittgenstein, *Remarks on the Philosophy of Psychology*, Chicago: Chicago University Press 1980.

Janet Wolff, 'The Invisible Flâneuse: Women and the Literature of Modernity' in Janet Wolff, *Feminine Sentences: Essays on Women and Culture*, Cambridge/Oxford: Polity Press 1990.

Janet Wolff, 'The artist and the flâneur: Rodin, Rilke and Gwen John in Paris' in Keith Tester (ed.), *The Flâneur*, London and New York: Routledge 1994.

Richard S.Wurman, *Information Architects*, Switzerland: Graphis Press Corp. 1996.

Index